International Association of Fire Chiefs

National Fire Protection Association

Fire Inspector

Principles and Practice

Revised First Edition

Student Workbook

JONES & BARTLETT LEARNING

Jones & Bartlett Learning
World Headquarters
5 Wall Street
Burlington, MA 01803
978-443-5000
info@jblearning.com
www.jblearning.com

International Association of Fire Chiefs
4025 Fair Ridge Drive
Fairfax, VA 22033
www.IAFC.org

National Fire Protection Association
1 Batterymarch Park
Quincy, MA 02169-7471
www.NFPA.org

Jones & Bartlett Learning books and products are available through most bookstores and online booksellers. To contact Jones & Bartlett Learning directly, call 800-832-0034, fax 978-443-8000, or visit our website, www.jblearning.com.

Substantial discounts on bulk quantities of Jones & Bartlett Learning publications are available to corporations, professional associations, and other qualified organizations. For details and specific discount information, contact the special sales department at Jones & Bartlett Learning via the above contact information or send an email to specialsales@jblearning.com.

Production Credits
Chief Executive Officer: Ty Field
President: James Homer
SVP, Editor-in-Chief: Michael Johnson
Executive Publisher: Kimberly Brophy
VP, Sales, Public Safety Group: Matthew Maniscalco
Director of Sales, Public Safety Group: Patricia Einstein
Executive Acquisitions Editor: William Larkin
Production Editor: Jessica deMartin
Senior Marketing Manager: Brian Rooney
VP, Manufacturing and Inventory Control: Therese Connell
Composition: Cenveo Publisher Services
Cover Design: Kristin E. Parker
Rights and Permissions Manager: Katherine Crighton
Photo Research Supervisor: Anna Genoese
Cover Image: © Dennis Wetherhold, Jr.
Printing and Binding: LSC Communications
Cover Printing: LSC Communications

ISBN: 978-1-284-14131-3

6048

Printed in the United States of America
21 20 19 18 17 10 9 8 7 6 5 4 3 2 1

Contents

Note to the reader: Exercises indicated with a [helmet icon] are specific to the Fire Inspector II level.

Student Resources

Navigate Test Prep: Fire Inspector I and II

ISBN: 978-0-7637-9860-4

Navigate Test Prep: Fire Inspector I and II is a dynamic program designed to prepare students to sit for Fire Inspector certification examinations by including the same type of questions they will likely see on the actual examination.

It provides a series of self-study modules, organized by chapter and level, offering practice examinations and simulated certification examinations using multiple-choice questions. All questions are page referenced to ***Fire Inspector: Principles and Practice*** for remediation to help students hone their knowledge of the subject matter.

Students can begin the task of studying for their fire inspector certification examinations by concentrating on those subject areas where they need the most help. Upon completion, students will feel confident and prepared to complete the final step in the certification process–passing the examination.

Exam Prep: Fire Inspector, Second Edition

ISBN: 978-1-4496-0964-1

Exam Prep: Fire Inspector, Second Edition is designed to thoroughly prepare you for Fire Inspector certification, promotion, or training examination by including the same type of multiple-choice questions you are likely to encounter on the actual exam.

To help improve examination scores, this prep guide follows the Performance Training Systems, Inc.'s Systematic Approach to Examination Preparation. ***Exam Prep: Fire Inspector, Second Edition*** was written by fire personnel explicitly for fire personnel, and all content was verified with the latest reference materials and by a technical review committee.

Your exam performance will improve after using this system!

Technology Resources

www.Fire.jbpub.com

This site has been specifically designed to complement ***Fire Inspector: Principles and Practice, Revised First Edition*** and is regularly updated. Resources available include:

- Chapter Pretests that prepare students for training. Each chapter has a pretest and provides instant results, feedback on incorrect answers, and page references for further study.
- Interactivities that allow students to reinforce their understanding of the most important concepts in each chapter.
- Hot Term Explorer, a virtual dictionary, allowing students to review key terms, test their knowledge of key terms through quizzes and flashcards, and complete crossword puzzles.

Introduction to Fire Inspector

The Workbook Activities

The following activities have been designed to help you. Your instructor may require you to complete some or all of these activities as a regular part of your fire inspector training program. You are encouraged to complete any activity that your instructor does not assign as a way to enhance your learning in the classroom.

Chapter Review

The following exercises provide an opportunity to refresh your knowledge of this chapter.

Matching

Match each of the terms in the left column to the appropriate definition in the right column.

_____	**1.** Fire Inspector I	**A.** Educates the public about fire safety and injury prevention and present juvenile fire safety programs.
_____	**2.** Fire inspection	**B.** Conducts most types of inspections and interprets applicable codes and standards.
_____	**3.** Fire marshals	**C.** Conducts basic fire inspections and applies codes and standards.
_____	**4.** Standard	**D.** Usually has an engineering degree, reviews plans, and works with building owners to ensure that their fire suppression and detection systems will meet code and function as needed.
_____	**5.** Fire Inspector II	**E.** A visual inspection of a building and its property that is conducted to determine if the building complies with all pertinent statutes and regulations of the jurisdiction.
_____	**6.** Fire and life safety education specialist	**F.** Inspects businesses and enforce public safety laws and fire codes and may respond to fire scenes to help investigate the cause of a fire.
_____	**7.** Prescriptive code	**G.** A rule or law established for enforcement of a fire protection, life safety, building construction, or property maintenance issue.
_____	**8.** Fire protection engineer	**H.** A document outlining the specifications or requirements defining the minimum levels of performance, protection, or construction.
_____	**9.** Permit	**I.** Lists the specific details the installation and construction must meet.
_____	**10.** Code	**J.** Document stating the compliance with or restrictions for use based on applicable codes.

CHAPTER 1

Multiple Choice

Read each item carefully, and then select the best response.

______ 1. Which of the following is *not* one of the three E's of fire prevention?
   A. Engineering
   B. Education
   C. Enforcement
   D. Entitlement

______ 2. Regulatory or industry-adopted documents that provide guidance for the safe construction and use of buildings are called:
   A. codes and standards.
   B. permits.
   C. building forms.
   D. fire engineering documents.

______ 3. A rule or law established for enforcement of a fire protection, life safety, building construction, or property maintenance issue is called a:
   A. permit.
   B. code.
   C. standard.
   D. regulation.

______ 4. A document outlining the specifications or requirements that define the minimum levels of performance, protection, or construction is a:
   A. permit.
   B. code.
   C. standard.
   D. regulation.

______ 5. This lists the specific details the installation and construction must meet, such as the type of electrical wiring to be used in an occupancy.
   A. National standard
   B. Performance-based code
   C. Prescriptive code
   D. Building rules

______ 6. This type of code does not state that a particular method or material must be used or design has to be met.
   A. National standard
   B. Performance-based code
   C. Prescriptive code
   D. Building rules

______ 7. Manufacturer- or industry-developed techniques to perform an activity in a safe manner are called:
   A. best options.
   B. standard practice.
   C. informal suggestions.
   D. recommended practice or guides.

_____ 8. The objective of the permit application process is to make certain that buildings:
- A. are safe for the public to occupy and fire fighters to operate.
- B. pay permit fees.
- C. meet occupancy requirements.
- D. are safe for the contractors during construction.

_____ 9. The key to improving ethical choices is to:
- A. follow the lead of the company officers.
- B. make choices based on what will help the inspector.
- C. have clear organizational values.
- D. make choices based on what will aid the building owner.

_____ 10. The authority to perform fire inspections is given by whom?
- A. The president
- B. The governor
- C. It differs from state to state
- D. The chief

Fill in the Blank

Read each item carefully, and then complete the statement by filling in the missing word(s).

1. A __________ __________ includes testing fire suppression, fire alarm, and detection systems, conducting occupant evacuation drills, evaluating the fire-retardant quality of materials used within the building, and providing guidelines to the building owners on fire-safe materials.
2. __________ __________ may have full police powers to investigate and arrest suspected arsonists and people causing false alarms.
3. According to NFPA 1031, a __________ __________ __________ conducts most types of inspections and interprets applicable codes and standards.
4. The __________ __________ __________ is intended to keep the people of the community safe—at work, at home, and when out and about in the community.
5. Local, county, district, or state officials may __________ __________ __________ the codes or standards that meet the specific needs of the jurisdiction.
6. The __________ __________ of the fire inspector should list the limits of legal authority that the position holds in the local jurisdiction.
7. To avoid legal disputes, a standardized __________ __________ __________ that has been reviewed for legal sufficiency by the attorney for the jurisdiction should be used by all personnel.
8. A __________ is a document stating the compliance with or restrictions for use based on applicable codes.
9. __________ __________ are based on a value system.

True/False

If you believe the statement to be more true than false, write the letter "T" in the space provided. If you believe the statement to be more false than true, write the letter "F."

_____ 1. Effective fire inspections, plan reviews, and public education efforts conducted by fire inspectors can prevent fires and the loss of life.

_____ 2. Eliminating is one of the "Three E's of Fire Prevention."

_____ 3. In addition to standard fire inspections, general fire safety or process inspections are also performed to check for overcrowding, blocked exits during business hours, or other types of unsafe practices that may be occurring while the building or space is occupied, placing the lives of the occupants at risk.

_______ 4. In some communities, local fire department engine or ladder companies continue to conduct fire inspections in addition to preplanning inspections.

_______ 5. A fire inspection is a visual inspection of a building and its property conducted to determine if the building complies with all pertinent statutes and regulations of the jurisdiction.

_______ 6. A Fire Inspector I is not trained to identify the occupancy classification of a single-use occupancy.

_______ 7. A Fire Inspector I may be required to participate in legal proceedings and provide testimony or written comments as required.

_______ 8. According to NFPA 1031, a Fire Inspector II should have the ability to classify the occupancy type of a building based on a set of plans, specifications, and a description of a building.

_______ 9. A standard may or may not be a legally mandated document, but is considered an industry best practice to follow.

_______ 10. The range of authority of the fire inspector is defined by federal standards.

Vocabulary

Define the following terms using the space provided.

1. Code:

____

____

____

2. Standard:

____

____

____

3. Fire marshal:

____

____

____

4. Fire Inspector I:

____

____

____

Short Answer

Complete this section with short written answers, using the space provided.

1. Identify six job performance requirements a Fire Inspector I must meet, per NFPA 1031.

____

____

2. Identify six job performance requirements a Fire Inspector II must meet, per NFPA 1031.

3. Identify four of the more common fire positions a fire inspector may also assume.

4. What path does the adoption process of codes and standards typically follow?

5. Many fire inspection activities result in the issuance of a permit. Identify four types of permits.

Fire Alarm

The following case scenario will give you an opportunity to explore the concerns associated with being a fire inspector. Read the scenario, and then answer each question in detail.

You are a recently appointed fire inspector. You are inspecting the local mayor's family business and have noticed several occupancy violations and various safety issues. The building manager, the mayor's brother, tells you the last inspector ignored these issues and never cited the business or required any changes. You stand by your own code of ethics and inform the manager he will have to comply with the code or you will cite him for the violations you have observed.

1. Within your department, how will you address the lack of organizational ethics, meaning the past act of allowing these violations to continue?

2. What is your legal authority and your range of authority in the scenario just described?

Building Construction

The Workbook Activities

The following activities have been designed to help you. Your instructor may require you to complete some or all of these activities as a regular part of your fire inspector training program. You are encouraged to complete any activity that your instructor does not assign as a way to enhance your learning in the classroom.

Chapter Review

The following exercises provide an opportunity to refresh your knowledge of this chapter.

Matching

Match each of the terms in the left column to the appropriate definition in the right column.

_____	**1.** Platform-frame construction	**A.**	An older type of wood frame construction in which the wall studs extend vertically from the basement of a structure to the roof without any fire stops.
_____	**2.** Truss	**B.**	Trusses that are curved on the top and straight on the bottom.
_____	**3.** Combustibility	**C.**	Glass that is much stronger and harder to break than ordinary glass.
_____	**4.** Fire partition	**D.**	A property that describes how quickly a material will conduct heat.
_____	**5.** Balloon-frame construction	**E.**	A collection of lightweight structural components joined in a triangular configuration that can be used to support either floors or roofs.
_____	**6.** Pitched chord truss	**F.**	Construction technique for building the frame of the structure one floor at a time. Each floor has a top and bottom plate that acts as a fire stop.
_____	**7.** Dead load	**G.**	The property describing whether a material will burn and how quickly it will burn.
_____	**8.** Thermal conductivity	**H.**	An interior wall extending from the floor to the underside of the floor above.
_____	**9.** Tempered glass	**I.**	Type of truss typically used to support a sloping roof.
_____	**10.** Bowstring trusses	**J.**	The weight of a building. It consists of the weight of all materials of construction incorporated into a building, including but not limited to walls, floors, roofs, ceilings, stairways, built-in partitions, finishes, cladding, and other similarly incorporated architectural and structural items, as well as fixed service equipment, including the weight of cranes.

Multiple Choice

Read each item carefully, and then select the best response.

_____ **1.** A roof that has sloping or inclined surfaces is called:

A. curved roof.
B. flat roof.
C. pitched roof.
D. suspended roof.

CHAPTER 2

_____ 2. Solid wood joists mounted in an inclined position are called:
A. bowstrings.
B. rafters.
C. trusses.
D. girder.

_____ 3. What type of floor is common in fire-resistive construction?
A. Wood
B. Laminate
C. Steel
D. Concrete floors

_____ 4. What type of construction is the most common type of construction used today?
A. Type V
B. Type IV
C. Type III
D. Type II

_____ 5. What type construction is the most fire-resistive category of building construction?
A. Type I
B. Type II
C. Type III
D. Type IV

_____ 6. What type construction is also referred to as ordinary construction because it is used in a wide variety of buildings, ranging from commercial strip malls to small apartment buildings?
A. Type I
B. Type II
C. Type III
D. Type V

_____ 7. What type construction is also known as heavy timber construction?
A. Type I
B. Type II
C. Type III
D. Type IV

_____ 8. Which of the following is *not* a variation of a pitched roof?
A. Gable
B. Hip
C. Mansard
D. Bowstring

_____ 9. What is a structural component composed of smaller components in a triangular configuration or a system of triangles?
A. Rafter
B. Truss
C. Beam
D. Joist

_____ 10. What are interior walls that extend from a floor to the underside of the floor above? They often enclose fire-rated interior corridors or divide a floor area into separate fire compartments.
- A. Fire partitions
- B. Party walls
- C. Nonbearing walls
- D. Bearing walls

_____ 11. What is the transparent part of a window most commonly called?
- A. Plate
- B. Annealed
- C. Glazed
- D. Tempered

_____ 12. What is the generic name for a family of sheet products consisting of a noncombustible core primarily of gypsum with paper surfacing?
- A. Gypsum board
- B. Hollow core
- C. Curtain wall
- D. Fire partition

Fill in the Blank

Read each item carefully, and then complete the statement by filling in the missing word(s).

1. _____ _____ are nonbearing exterior walls attached to the outside of the building.
2. _____ _____ are interior nonbearing walls that can easily be removed and replaced.
3. _____ _____ extend above the roof line of a building.
4. A _____ is a beam that supports other beams.
5. A _____ is a beam specifically used to support a roof system.
6. Fire inspectors must constantly determine the _____ of a structural element relative to the overall stability of the building.
7. The _____ of the foundation system, load-bearing walls and vertical columns is critical to the structural stability of all buildings.
8. Single-pane glass, also called regular or _____ glass, is normally used in construction because it is relatively inexpensive.
9. Wired glass is tempered glass that has been reinforced with wire. This kind of glass may be clear or frosted, and it is often used in _____ _____ doors.
10. A _____ window is made of adjustable sections of tempered glass encased in a metal frame that overlap each other when closed.

True/False

If you believe the statement to be more true than false, write the letter "T" in the space provided. If you believe the statement to be more false than true, write the letter "F."

_____ 1. From a structural standpoint, the ceiling is to be considered part of the floor assembly for the story above.

_____ 2. Awning windows are similar in operation to double-hung windows, except that they usually have one large or two medium-size glass panels.

_____ 3. Fire doors and fire windows come in many different shapes and sizes, and they provide different levels of fire resistance.

_____ 4. All fire doors must have a mechanism that keeps the door closed or automatically closes the door when a fire occurs.

_____ 5. A fire door rated at 1 hour will probably last half as long as a fire door rated 30 minutes.

_____ 6. Fire doors and windows are labeled and assigned the letters A, B, C, D, or E, based on their approved-use locations.

_____ 7. Glass blocks do not resist high temperatures and frequently melt down during a structure fire.

_____ 8. Different interior finish materials contribute in various ways to a building fire.

_____ 9. Buildings that are under construction and buildings that are being renovated or demolished are at high risk for major fires.

_____ 10. Hollow-core doors are difficult to force and will not burn through quickly.

Vocabulary

Define the following terms using the space provided.

1. Pyrolysis:

_____

_____

_____

2. Parapet wall:

_____

_____

_____

3. Spalling:

_____

_____

_____

4. Rafters:

_____

_____

_____

5. Dead load:

_____

_____

_____

Short Answer

Complete this section with short written answers, using the space provided.

1. List the eight most common building materials.

2. Identify the four key factors that affect the behavior of these common building materials under fire conditions.

3. Identify five different types of wood products commonly used in building applications today.

4. Identify and briefly describe five types of building construction.

5. Identify and briefly describe the five NFPA 80 designations for fire doors and fire windows.

Fire Alarm

The following case scenario will give you an opportunity to explore the concerns associated with being a fire inspector. Read the scenario and then answer each question in detail.

As the sole building inspector in your jurisdiction, you are occasionally called in to assist incident commanders in large structure fires. You have just been called in to a multi-alarm fire in a construction type IV structure that has been converted from a granary to a retail outlet. The fire has been burning for several hours.

1. What advice could you give to the incident commander concerning fire characteristics in this type of structure?

2. What is the probable outcome of this fire in this type of structure, considering the duration of the burn?

Types of Occupancies

The Workbook Activities

The following activities have been designed to help you. Your instructor may require you to complete some or all of these activities as a regular part of your fire inspector training program. You are encouraged to complete any activity that your instructor does not assign as a way to enhance your learning in the classroom.

Chapter Review

The following exercises provide an opportunity to refresh your knowledge of this chapter.

Matching

Match each of the terms in the left column to the appropriate definition in the right column.

_____	**1.** Occupancy	**A.**	A building or portion thereof containing three or more dwelling units with independent cooking and bathroom facilities.
_____	**2.** Dormitory	**B.**	Buildings (1) used for a gathering of 50 or more persons for deliberation, worship, entertainment, eating, drinking, amusement, awaiting transportation, or similar uses; or (2) used as a special amusement building regardless of occupant load.
_____	**3.** Mixed occupancy	**C.**	Buildings used for educational purposes through the twelfth grade by six or more persons for 4 or more hours per day or more than 12 hours a week.
_____	**4.** Detention and correctional occupancy	**D.**	A building or space in a building in which group sleeping accommodations are provided for more than 16 persons who are not members of the same family in one room, or a series of closely associated rooms, under joint occupancy and single management, with or without meals, but without individual cooking facilities.
_____	**5.** Daycare occupancy	**E.**	A building that contains no more than two dwelling units with independent cooking and bathroom facilities.
_____	**6.** Mercantile occupancy	**F.**	An occupancy in which four or more clients receive care, maintenance, and supervision, by other than their relatives or legal guardians, for less than 24 hours per day.
_____	**7.** Assembly occupancies	**G.**	A multiple occupancy where the occupancies are intermingled.
_____	**8.** Apartment building	**H.**	An occupancy used to one or more persons under varied degrees of restraint or security where such occupants are mostly incapable of self-preservation because of security measures not under the occupant's control.
_____	**9.** Educational occupancies	**I.**	The intended use of a building
_____	**10.** One- or two-family dwelling	**J.**	An occupancy used for the display and sale of merchandise.

CHAPTER

3

Multiple Choice

Read each item carefully, and then select the best response.

______ 1. A building or portion thereof used to provide services or treatment simultaneously to four or more patients that, on an outpatient basis, is a(n):
  - A. mixed occupancy.
  - B. ambulatory healthcare occupancy.
  - C. daycare occupancy.
  - D. business occupancy.

______ 2. An occupancy in which products are manufactured or in which processing, assembling, mixing, packaging, finishing, decorating, or repair operations are conducted is a(n):
  - A. daycare occupancy.
  - B. business occupancy.
  - C. industrial occupancy.
  - D. storage occupancy.

______ 3. Building or portion thereof that does not qualify as a one- or two-family dwelling, that provides sleeping accommodations for a total of 16 or fewer people on a transient or permanent basis, without personal care services, with or without meals, but without separate cooking facilities for individual occupants is a(n):
  - A. lodging or rooming house.
  - B. hotel.
  - C. ambulatory healthcare occupancy.
  - D. multiple occupancy.

______ 4. An occupancy used for the transaction of business other than mercantile is called a:
  - A. multiple occupancy.
  - B. business occupancy.
  - C. storage occupancy.
  - D. mixed occupancy.

______ 5. A multiple occupancy where the occupancies are separated by fire resistance rated assemblies is a:
  - A. multiple occupancy.
  - B. mixed occupancy.
  - C. storage occupancy.
  - D. separated occupancy.

______ 6. An occupancy used primarily for the storage or sheltering of goods, merchandise, products, and vehicles is a(n):
  - A. storage occupancy.
  - B. separated occupancy.
  - C. multiple occupancy.
  - D. industrial occupancy.

_____ 7. A building or portion thereof that is used for lodging and boarding of four or more residents, not related by blood or marriage to the owners or operators, for the purpose of providing personal care services is a:
- **A.** dormitory.
- **B.** hotel.
- **C.** residential board and care occupancy.
- **D.** daycare.

_____ 8. An occupancy used for purposes of medical or other treatment or care of four or more persons where such occupants are mostly incapable of self-preservation due to age, physical or mental disability, or because of security measures not under the occupant's control is called a(n):
- **A.** detention and correctional occupancy.
- **B.** residential board and care occupancy.
- **C.** daycare occupancy.
- **D.** healthcare occupancy.

_____ 9. Building or portion thereof that does not qualify as a one- or two-family dwelling, that provides sleeping accommodations for a total of 16 or fewer people on a transient or permanent basis, without personal care services, with or without meals, but without separate cooking facilities for individual occupants is a(n):
- **A.** dormitory.
- **B.** lodging or rooming house.
- **C.** hotel.
- **D.** apartment building.

_____ 10. A building or structure in which two or more classes of occupancy exist is a:
- **A.** dormitory.
- **B.** lodging or rooming house.
- **C.** multiple occupancy.
- **D.** hotel.

_____ 11. A building or space in a building in which group sleeping accommodations are provided for more than 16 persons who are not members of the same family is a(n):
- **A.** dormitory.
- **B.** lodging or rooming house.
- **C.** hotel.
- **D.** apartment building.

_____ 12. An occupancy used for the transaction of business other than mercantile is what type of occupancy?
- **A.** Industrial occupancy
- **B.** Business occupancy
- **C.** Storage occupancy
- **D.** Nonmercantile occupancy

Fill in the Blank

Read each item carefully, and then complete the statement by filling in the missing word(s).

1. In the model building codes, the term *occupancy* refers to the ____________ use of a building.
2. Because a building can change ____________ ____________ over time, it is critical that you correctly identify the current use of the building and the appropriate occupancy classification under the applicable building codes and regulations.
3. An incorrect occupancy classification could result in requiring fire protection systems that are ____________ ____________ for the occupancy.
4. In conducting an inspection of a one- or two-family home, you typically focus on the means of escape, the interior finishes, and ____________ ____________.
5. Of particular concern as you inspect ____________ ____________ will be the building compartmentation and the provision of adequate escape routes.

6. Based on the adoption of new codes and requirements over time, the code requirements are often ________________ for existing apartment buildings and new apartment buildings.

7. If nursing care is provided, then the facility is considered to be a(n) ________________ ________________; however, if nursing care is not provided, it is considered to be a residential board and care occupancy.

8. A healthcare facility used only for outpatients is addressed as a(n) ________________ healthcare occupancy.

9. The three subclasses within the ________________ occupancy category are distinguished based on the number of clients serviced.

10. When inspecting a(n) ________________ occupancy, you should consider occupant load, means of egress, interior finish, protection of openings, and protection from hazards.

True/False

If you believe the statement to be more true than false, write the letter "T" in the space provided. If you believe the statement to be more false than true, write the letter "F."

______ 1. A one- or two-family dwelling is a building that contains no more than two dwelling units with independent cooking and bathroom facilities.

______ 2. A hotel is a building or portion thereof, not categorized as a one- or two-family dwelling, that provides sleeping accommodations for a total of 16 or fewer people on a transient or permanent basis.

______ 3. A lodging or rooming house is a building or space in a building in which group sleeping accommodations are provided for more than 16 persons who are not members of the same family.

______ 4. A residential board and care occupancy is a building or portion thereof that is used for lodging and boarding of four or more residents, not related by blood or marriage to the owners or operators, for the purpose of providing personal care services.

______ 5. A healthcare occupancy is an occupancy used for purposes of medical or other treatment or care of four or more persons on an inpatient basis where such occupants are mostly incapable of self-preservation due to age, physical or mental disability, or because of security measures not under the occupant's control.

______ 6. An ambulatory healthcare occupancy is an occupancy used to provide outpatient services or treatment to four or more patients simultaneously.

______ 7. An industrial occupancy is an occupancy used for the transaction of business other than mercantile.

______ 8. A business occupancy is an occupancy used for the display and sale of merchandise. Included under this occupancy classification are shopping centers, supermarkets, drugstores, department stores, auction rooms, restaurants with fewer than 50 persons, and any occupancy or portion thereof that is used for the display and sale of merchandise.

______ 9. A storage occupancy is an occupancy used primarily for the storage or sheltering of goods, merchandise, products, and vehicles.

______ 10. An educational occupancy is an occupancy used to house one or more persons under varied degrees of restraint or security where such occupants are mostly incapable of self-preservation because of security measures not under the occupant's control.

Vocabulary

Define the following terms using the space provided.

1. Occupancy:

2. Mixed occupancy:

3. Multiple occupancy:

Short Answer

Complete this section with short written answers, using the space provided.

1. List the 15 specific occupancy groupings, according to NFPA 101.

2. List six examples of residential board and care occupancies.

3. List three examples of services provided by ambulatory healthcare occupancies.

4. Identify three subclasses of daycare occupancies.

5. Identify and briefly describe the five categories under detention and correctional occupancy that correspond to the degree of restraint of occupants within the facility.

Fire Alarm

The following case scenario will give you an opportunity to explore the concerns associated with being a fire inspector. Read the scenario, and then answer each question in detail.

A group of developers has built a 175-foot (53.3-m) concrete tower with a restaurant at the top. This structure does not meet a specific occupancy classification due to its construction and multiple uses (it also has a large conference space at ground level, as well as some mercantile shops).

1. How will you determine what requirements the safeguards of this mixed occupancy building should comply with?

2. How will you determine the occupancy classification and any special requirements that apply to this structure?

____

____

____

____

____

____

____

____

____

____

Fire Growth

The Workbook Activities

The following activities have been designed to help you. Your instructor may require you to complete some or all of these activities as a regular part of your fire inspector training program. You are encouraged to complete any activity that your instructor does not assign as a way to enhance your learning in the classroom.

Chapter Review

The following exercises provide an opportunity to refresh your knowledge of this chapter.

Matching

Match each of the terms in the left column to the appropriate definition in the right column.

______	**1.** Backdraft	**A.** A geometric shape used to depict the four components required for a fire to occur: fuel, oxygen, heat, and chemical chain reactions.
______	**2.** Convection	**B.** A rapid, persistent chemical reaction that releases both heat and light.
______	**3.** Flash point	**C.** Heat transfer to another body or within a body by direct contact.
______	**4.** Upper explosive limit (UEL)	**D.** The sudden explosive ignition of fire gases when oxygen is introduced into a superheated space previously deprived of oxygen.
______	**5.** Fire tetrahedron	**E.** All combustible materials. The actual material that is being consumed by a fire, allowing the fire to take place.
______	**6.** Fire	**F.** The upper and lower concentration limits (at a specified temperature and pressure) of a flammable gas or vapor in air that can be ignited, expressed as a percentage of the fuel by volume.
______	**7.** Fuel	**G.** The minimum amount of gaseous fuel that must be present in the air mixture for the mixture to be flammable or explosive.
______	**8.** Conduction	**H.** Heat transfer by circulation within a medium such as a gas or a liquid.
______	**9.** Lower explosive limit (LEL)	**I.** The maximum amount of gaseous fuel that can be present in the air mixture for the mixture to be flammable or explosive.
______	**10.** Flammability limits (explosive limits)	**J.** The minimum temperature at which a liquid or a solid releases sufficient vapor to form an ignitable mixture with the air.

Multiple Choice

Read each item carefully, and then select the best response.

______ **1.** Fires that involve energized electrical equipment, where the electrical conductivity of the extinguishing media is of importance are:

A. Class A.
B. Class B.
C. Class C.
D. Class D.
E. Class K.

CHAPTER

4

_____ **2.** Fires involving flammable and combustible liquids, oils, greases, tars, oil-based paints, lacquers, and flammable gases are:
- **A.** Class A.
- **B.** Class B.
- **C.** Class C.
- **D.** Class D.
- **E.** Class K.

_____ **3.** Fires involving ordinary combustible materials, such as wood, cloth, paper, rubber, and many plastics are:
- **A.** Class A.
- **B.** Class B.
- **C.** Class C.
- **D.** Class D.
- **E.** Class K.

_____ **4.** Fires involving combustible cooking media such as vegetable oils, animal oils, and fats are:
- **A.** Class A.
- **B.** Class B.
- **C.** Class C.
- **D.** Class D.
- **E.** Class K.

_____ **5.** Fires involving combustible metals such as magnesium, titanium, zirconium, sodium, and potassium are:
- **A.** Class A.
- **B.** Class B.
- **C.** Class C.
- **D.** Class D.
- **E.** Class K.

_____ **6.** Energy that is created or released by the combination or decomposition of chemical compounds is:
- **A.** electrical energy.
- **B.** chemical energy.
- **C.** mechanical energy.
- **D.** volatility.

_____ **7.** Heat that is produced by electricity is:
- **A.** electrical energy.
- **B.** chemical energy.
- **C.** mechanical energy.
- **D.** volatility.

_____ **8.** Heat that is created by friction is:
- **A.** electrical energy.
- **B.** chemical energy.
- **C.** mechanical energy.
- **D.** volatility.

_____ 9. The minimum temperature at which a fuel, when heated, ignites in air and continues to burn is:
   A. flashpoint.
   B. ignition phase.
   C. flameover.
   D. ignition temperature.

_____ 10. A condition in which all combustibles in a room or confined space have been heated to the point at which they release vapors that will support combustion, causing all combustibles to ignite simultaneously is:
   A. growth phase.
   B. flashover.
   C. decay phase.
   D. back draft.

_____ 11. These assume the shape of the container in which they are placed, and for all practical purposes do not compress.
   A. Solids
   B. Liquids
   C. Gases
   D. Other

_____ 12. Hot burning gases traveling horizontally along a ceiling are commonly referred to as:
   A. conduction currents.
   B. plume or mushroom.
   C. thermal column.
   D. rollover or flameover.

Fill in the Blank

Read each item carefully, and then complete the statement by filling in the missing word(s).

1. The key to preventing a __________ is to cool the top of the tank that contains the vapor.
2. During the __________ __________, the rate of burning slows down because less fuel is available.
3. To understand the behavior of fire, you need to consider the three basic elements needed for a fire to occur: fuel, oxygen, and heat. If we graphically place these three together, the result is a __________ __________.
4. __________ is the flaming ignition of hot gases that are layered in a developing room or compartment fire.
5. The __________ __________ is the lowest temperature at which a liquid produces enough vapor to sustain a continuous fire.
6. The __________ __________ phase lasts as long as a large supply of fuel is available.
7. A __________ is a type of liquid that has neither independent shape nor independent volume but rather tends to expand indefinitely.
8. The __________ __________, when the fire spreads to nearby fuel, occurs as the kindling starts to burn, increasing the convection of hot gases upward.
9. __________ is a state of inadequate oxygenation of the blood and tissue.
10. __________ is the transfer of heat through the emission of energy in the form of invisible waves.

True/False

If you believe the statement to be more true than false, write the letter "T" in the space provided. If you believe the statement to be more false than true, write the letter "F."

_____ 1. Endothermic reactions absorb heat or require heat to be added.

_____ 2. Exothermic reactions result in the release of energy in the form of heat.

_____ 3. A solid is nongaseous substance composed of molecules that move and flow freely and assumes the shape of the container that holds it.

_____ 4. The ability of a substance to produce combustible vapors is its vapor density.

_____ 5. The stratification (heat layers) that occurs in a room as a result of a fire is called a plume.

_____ 6. A thermal column is a cylindrical area above a fire in which heated air and gases rise and travel upward.

_____ 7. A fuel is any material that stores potential energy.

_____ 8. Class C fires involve combustible cooking oils and fats in kitchens.

_____ 9. The growth phase, in which the fire is limited to its point of origin, begins as a lighted match is placed next to a crumpled piece of paper.

_____ 10. Fires burn without an adequate supply of oxygen, which results in incomplete combustion and produces a variety of toxic by-products, collectively called smoke.

Vocabulary

Define the following terms using the space provided.

1. Endothermic:

_____

_____

_____

2. Chemical energy:

_____

_____

_____

3. Fire tetrahedron:

_____

_____

_____

4. Vapor density:

_____

_____

_____

5. Volatility:

_____

_____

_____

Short Answer

Complete this section with short written answers, using the space provided.

1. Fires grow and spread by three primary mechanisms. Identify and briefly describe them.

2. Identify three toxic gases found in smoke.

3. Identify four methods of extinguishing a fire.

4. Identify and describe five classes of fire.

5. There are four solid-fuel fire phases. Identify and describe them.

Fire Alarm

The following case scenario will give you an opportunity to explore the concerns associated with being a fire inspector. Read the scenario, and then answer each question in detail.

Your fire chief has requested you attend a public function and prepare a brief presentation on residential fires and hidden building materials that may contribute to fire spread.

1. What hidden building materials and components have been known to contribute to fire growth?

2. How do interior finishes contribute to fire spread, and increase a fire hazard?

Performing an Inspection

The Workbook Activities

The following activities have been designed to help you. Your instructor may require you to complete some or all of these activities as a regular part of your fire inspector training program. You are encouraged to complete any activity that your instructor does not assign as a way to enhance your learning in the classroom.

Chapter Review

The following exercises provide an opportunity to refresh your knowledge of this chapter.

Matching

Match each of the terms in the left column to the appropriate definition in the right column.

	Term	Definition
_____	**1.** Annual inspections	**A.** Inspections that occur when the building department is notified of a new business requesting permission to open
_____	**2.** As-built diagrams	**B.** Is not used frequently and it must be used judicially. It should be used when contractors do not have the clearance for performing the work, or when the work is not correct and must be corrected prior to performing additional work
_____	**3.** Business license	**C.** Inspections performed as part of the regular inspection cycle
_____	**4.** Complaint inspections	**D.** Inspections that occur when someone registers a concern of a possible code violation
_____	**5.** Exigent circumstance	**E.** Set of drawings provided by a contractor showing how a system was actually installed, which may be different from the approved plans
_____	**6.** Change of occupancy inspections	**F.** Occur when code violations have been noted and you are checking to see if the owner is now compliant with the code
_____	**7.** Reinspections	**G.** Form that lists in detail any complaint that is lodged with the fire inspection agency and is investigated
_____	**8.** Complaint form	**H.** Those that are likely to burn with extreme rapidity and from which explosion is likely
_____	**9.** Stop work order	**I.** An immediate life safety issue that requires immediate actions to be taken
_____	**10.** High hazard contents	**J.** Developed through a consensus process

Multiple Choice

Read each item carefully, and then select the best response.

_____ **1.** These types of inspections occur when code violations have been noted and you are checking to see if the owner is now compliant with the code.

- **A.** Complaint inspections
- **B.** Final inspections
- **C.** Business license inspections
- **D.** Reinspections

CHAPTER

5

_____ 2. These type of inspections occur when someone registers a concern of a possible code violation.
   A. Complaint inspections
   B. Final inspections
   C. Business license inspections
   D. Reinspections

_____ 3. These are not performed by a fire inspector but by the business owner.
   A. Change of occupancy inspections
   B. Complaint inspections
   C. Self-inspections
   D. Conditional approvals

_____ 4. This is the law that gives you the right to note any violations of the code and have them corrected.
   A. Enabling legislation
   B. NFPA 1, 2009 edition
   C. NFPA 1, 2006 edition
   D. NFPA 220

_____ 5. These types of inspections ensure compliance with fire code during and at the end of building construction.
   A. Complaint inspections
   B. Final inspections
   C. Business license inspections
   D. Reinspections

_____ 6. This form lists in detail any complaint that is lodged with the fire inspection agency and is investigated.
   A. Inspection forms
   B. Final or construction inspection forms
   C. Complaint forms
   D. Stop work orders

_____ 7. All of the following should be part of your exterior building inspection except:
   A. fire hydrants.
   B. sprinkler connections.
   C. gates and fences.
   D. electric panel.

_____ 8. Buildings with content classified as those that are likely to burn with moderate rapidity and give off a considerable volume of smoke are:
   A. high hazard content.
   B. medium hazard content.
   C. low hazard content.
   D. ordinary hazard content.

_____ 9. Reinspection dates should be set:
   A. within 5 business days of the initial inspection.
   B. within 10 business days of the initial inspection.
   C. within 30 business days of the initial inspection.
   D. at the inspector's discretion.

_____ 10. These forms are used when inspecting specialized systems such as fire alarm, sprinkler, hood and duct suppression systems, as well as for other types of construction phase inspections.
- A. Inspection forms
- B. Final or construction inspection forms
- C. Complaint forms
- D. Stop work orders

_____ 11. This type of approval allows a business with minor nonhazardous violations to open, but requires that those violations be addressed.
- A. Conditional
- B. Temporary
- C. Reinspection
- D. Enabling

Fill in the Blank

Read each item carefully, and then complete the statement by filling in the missing word(s).

1. A fire inspection can reasonably ensure that a building will be __________ for the occupants.
2. __________ __________ __________ show how the final installation of equipment and utilities was actually completed.
3. It is critical to know which codes you can __________ enforce in your jurisdiction.
4. If inspecting a large complex, it may be helpful if you have a __________ __________ of the property.
5. When conducting the fire inspection, follow a __________ __________ __________ order so that no areas of the building are missed.
6. All __________ must be kept 36" (914 mm) away from a heat source.
7. Having the building and fire protection features meet the __________ __________ is a must.
8. As building plans are submitted there should be listing of what the __________ __________ are and rated for in the various areas within the building.
9. Carpets placed on walls lose their __________ __________ unless they have been tested and listed for wall use.
10. Oftentimes a __________ sketch for the fire department is drawn during the inspection process.

True/False

If you believe the statement to be more true than false, write the letter "T" in the space provided. If you believe the statement to be more false than true, write the letter "F."

_____ 1. Although the codes give you the authority to conduct fire inspections, you are not allowed to do so without permission of the owner.

_____ 2. If you have inspected a building before, it may be a good idea to rotate other fire inspectors through the buildings.

_____ 3. A kitchen may call for a specialized fire suppression system.

_____ 4. Elevators and escalators are intricate devices, and you should be involved in their actual testing.

_____ 5. When fire protection systems are installed you do not need to see the installation as it is progressing.

_____ 6. Cooking equipment generally does not require much in regard to special inspections.

_____ 7. From an inspection point of view, a remodeling of a building is no different than a new building.

_____ 8. Remodeling an existing building often activates requirements that the area being worked on, or in some cases the entire building, must be brought to the current codes for new construction.

_____ 9. It is not important to have a conversation with the owner regarding the findings of the inspection.

_____ 10. When responding to a complaint, the owner should not know that you plan to inspect because it is important that you see the condition as it exists not after there has been an opportunity to repair it.

Vocabulary

Define the following terms using the space provided.

1. Business license or change of occupancy inspections:

_____

2. Complaint inspections:

_____

3. Complaint form:

_____

Short Answer

Complete this section with short written answers, using the space provided.

1. List the elements that should be included on an inspection form.

_____

2. Identify some of the more common housekeeping safety issues.

3. Identify and briefly describe eight of the more common hazard violations.

4. Identify at least 10 building features you should inspect.

5. Identify six routine inspections a fire inspector may be required to perform.

Fire Alarm

The following case scenario will give you an opportunity to explore the concerns associated with being a fire inspector. Read the scenario, and then answer each question in detail.

You have recently completed your fire inspector certification course and are ready to hit the line. However, you feel uncomfortable in your limited knowledge and realize you have virtually no experience.

1. Identify some avenues available to you that may help you gather more knowledge about your new job.

2. How can you gain and then increase your experience level?

Reading Plans

The Workbook Activities

The following activities have been designed to help you. Your instructor may require you to complete some or all of these activities as a regular part of your fire inspector training program. You are encouraged to complete any activity that your instructor does not assign as a way to enhance your learning in the classroom.

Chapter Review

The following exercises provide an opportunity to refresh your knowledge of this chapter.

Matching

Match each of the terms in the left column to the appropriate definition in the right column.

	Term	Definition
______	**1.** Plan set	**A.** A design process whose fire safety solutions are designed to achieve a specified goal for a specified use or application.
______	**2.** Fire protection engineer	**B.** The use of systems, methods, or devices of equivalent or superior quality, strength, fire resistance, effectiveness, and durability to those prescribed by a code or standard.
______	**3.** Electrical plans	**C.** This clause allows for the code provisions to be altered and an alternative offered that would not reduce the level of safety within the building.
______	**4.** Equivalencies	**D.** A summary of the features of fire protection and building characteristics in a plan set.
______	**5.** Plan view	**E.** A collection of all of the information about a project that may be provided to the fire inspector in addition to the plan set during a plan review.
______	**6.** Architectural plan	**F.** Created by design professionals, plan sets include a series of drawings detailing how a proposed building will be built. Also known as plans, blueprints, construction documents, or shop drawings.
______	**7.** Specifications book	**G.** An engineer with specialized training in the protection of life and property from fire events.
______	**8.** Alternative clause	**H.** A drawing showing floor plans, elevation drawings, and features of a proposed building's layout and construction.
______	**9.** Code analysis	**I.** Design documents in a plan set showing the power layout and lighting plan of a proposed building.
______	**10.** Performance-based design	**J.** A view on a drawing where a horizontal slice is made in the building or area and everything above or below the slice is shown.

CHAPTER 6

Multiple Choice

Read each item carefully, and then select the best response.

_______ **1.** A plan set view where a horizontal slice is made in the building or area and everything above or below the slice is shown is a:
- **A.** plan view.
- **B.** elevation view.
- **C.** sectional view.
- **D.** detail view.

_______ **2.** The plan review process phase where you will examine all of the plans contained in the plan set is the:
- **A.** application phase.
- **B.** review phase.
- **C.** approval phase.
- **D.** commissioning phase.

_______ **3.** A plan set view that shows the exterior of the building and that will be labeled either using the direction the drawing is facing or by front, rear, left, and right is the:
- **A.** plan view.
- **B.** elevation view.
- **C.** sectional view.
- **D.** detail view.

_______ **4.** The plan review process phase to be completed after all of the plans in the plan set meet your local codes and standards is the:
- **A.** application phase.
- **B.** review phase.
- **C.** approval phase.
- **D.** commissioning phase.

_______ **5.** A vertical slice of a building showing the internal view is the:
- **A.** plan view.
- **B.** elevation view.
- **C.** sectional view.
- **D.** detail view.

_______ **6.** Generally the first look at the proposed building in relation to lot lines and other structures is with which plan?
- **A.** Structural plan
- **B.** Site plan
- **C.** Architectural plan
- **D.** Master plan

_______ **7.** Views of a specific element of construction or building feature in a larger scale providing more clarity are:
- **A.** plan views.
- **B.** elevation views.
- **C.** sectional views.
- **D.** detail views.

_____ 8. These plans contain information regarding the building's load-bearing components.
- A. Structural plan
- B. Site plan
- C. Architectural plan
- D. Master plan

_____ 9. These plans are where a majority of the information for compliance with the adopted codes is located.
- A. Structural plan
- B. Site plan
- C. Architectural plan
- D. Master plan

_____ 10. The final step in the plan review process consists of a final inspection and what other task?
- A. Plan rewrite
- B. Commissioning systems
- C. Final review
- D. Site visit

Fill in the Blank

Read each item carefully, and then complete the statement by filling in the missing word(s).

1. A successful building design is not complete until it is __________ according to the approved plan set.
2. *Plans, blueprints, construction documents, shop drawings*, and *plan sets* are all terms used to refer to the documents that will be reviewed during a __________ __________.
3. __________ __________ will be the first look at a proposed structure and will allow you to determine the construction type if it is not listed in the notes or summary.
4. The __________ __________ will also show any hydrants that will be installed on the site, and you should use this to determine if hydrant spacing is appropriate for the structure.
5. The first thing that you should look for when reviewing the architectural plans is the __________ classification.
6. It is important to ensure that when a fire-resistance rated assembly is penetrated with a duct for ventilation that the duct is provided with a __________ that prevents the spread of fire into that shaft.
7. The electrical and mechanical plans show information about the building utilities and __________ systems.
8. The adopted model building code and *Life Safety Code* are the driving force in the requirements for __________.
9. The first thing to look at with a fire __________ plan is the intent of the system.
10. When __________ are found during a plan review, they need to be brought to the attention of the submitter and, depending on the severity, may require correction before the permit is issued.

True/False

If you believe the statement to be more true than false, write the letter "T" in the space provided. If you believe the statement to be more false than true, write the letter "F."

_____ 1. A successful building design is not complete until it is built according to the approved plan set.

_____ 2. As the fire inspector, you are also ensuring that fire protection systems will operate as designed and achieve their performance goals.

_____ 3. As a rule, all documents marked "not for construction" should be reviewed by fire inspectors.

_____ 4. Smaller projects may not be drawn to scale, and as long as there is sufficient dimensioning to determine that code requirements are met, this should not be an issue.

_______ 5. There are four commonly used types of drawing scales: architect's scale, engineer's scale, mechanic's scale, and NFPA scale.

_______ 6. The site plan is a drawing showing the proposed building's load-bearing components.

_______ 7. The specifications book is a collection of all of the information about a certain project, including the details of the components required to meet the project's design criteria.

_______ 8. Electrical plans show the heating, ventilation, and air-conditioning layouts of the proposed building.

_______ 9. The class of fire extinguisher required depends on how the contents of the building are classified.

_______ 10. Most of the model codes and standards have provisions for equivalencies and alternatives.

Vocabulary

Define the following terms using the space provided.

1. Architectural plan:

____

____

____

2. Mechanical plans:

____

____

____

3. Plan view:

____

____

____

4. Sectional view:

____

____

____

5. Site plan:

____

____

____

Short Answer

Complete this section with short written answers, using the space provided.

1. Most drawings in plan sets contain a title block that is normally on the right side of the page. The title block contains what basic information about the drawing?

2. The code analysis should list the applicable building and fire codes for the project and based on those codes should show what information?

3. List and give a basic overview of the drawing types that you will encounter in a plan set.

4. There are four basic types of views that you will encounter in a plan set. List and give a brief description of each.

5. Most of the model codes and standards have provisions for equivalencies and alternatives. Describe what alternative clauses and equivalencies do.

Fire Alarm

The following case scenario will give you an opportunity to explore the concerns associated with being a fire inspector. Read the scenario and then answer each question in detail.

You are field inspecting a new construction site. After walking through the site several times and meeting with the construction foreman, you realize the approved plan set is not being followed.

1. What action should you take to remedy this situation?

After many meetings and discussions with the builders and engineers you begin to believe this new construction is beyond your knowledge base and capabilities.

2. What resources are available to assist you with this assignment?

Occupancy Safety and Evacuation Plans

The Workbook Activities

The following activities have been designed to help you. Your instructor may require you to complete some or all of these activities as a regular part of your fire inspector training program. You are encouraged to complete any activity that your instructor does not assign as a way to enhance your learning in the classroom.

Chapter Review

The following exercises provide an opportunity to refresh your knowledge of this chapter.

Matching

Match each of the terms in the left column to the appropriate definition in the right column.

	Term	Definition
_____	**1.** Area of refuge	**A.** An exit between adjacent areas on the same deck that passes through an A-60 Class boundary that is contiguous from side shell to side shell or to other A-60 Class boundaries.
_____	**2.** Panic hardware	**B.** A passageway from which there is only one means of egress.
_____	**3.** Smoke-proof enclosures	**C.** An area that is either (1) a story in a building where the building is protected throughout by an approved, supervised automatic-sprinkler system and has not less than two accessible rooms or spaces separated from each other by smoke-resisting partitions, or (2) a space located in a path of travel leading to a public way that is protected from the effects of fire, either by means of separation from other spaces in the same building or by virtue of location, thereby permitting a delay in egress travel from any level.
_____	**4.** Common path of travel	**D.** A stair enclosure designed to limit the movement of products of combustion produced by a fire.
_____	**5.** Dead-end corridor	**E.** A continuous and unobstructed way of exit travel from any point in a building or structure to a public way, consisting of three separate and distinct parts: (1) the exit access, (2) the exit, and (3) the exit discharge. A means of egress comprises the vertical and horizontal travel and includes intervening room spaces, doorways, hallways, corridors, passageways, balconies, ramps, stairs, enclosures, lobbies, escalators, horizontal exits, courts, and yards.
_____	**6.** Exit	**F.** A door-latching assembly incorporating a device that releases the latch upon the application of a force in the direction of egress travel.
_____	**7.** Exit access	**G.** That portion of a means of egress that is separated from all other spaces of a building or structure by construction or equipment as required to provide a protected way of travel to the exit discharge.
_____	**8.** Horizontal exit	**H.** That portion of a means of egress that leads to an exit.
_____	**9.** Exit discharge	**I.** That portion of a means of egress between the termination of an exit and a public way.
_____	**10.** Means of egress	**J.** The portion of exit access that must be traversed before two separate and distinct paths of travel to two exits are available.

CHAPTER

7

Multiple Choice

Read each item carefully, and then select the best response.

_______ 1. The *Life Safety Code* forms the basis for most of the egress requirements contained in the model building and fire codes. The *Life Safety Code* is:
A. NFPA 101.
B. NFPA 1043.
C. NFPA 1500.
D. NFPA 1021.

_______ 2. The gross occupant load factor is calculated for:
A. the space that can be occupied.
B. the building as a whole and is measured from wall to wall.
C. both the occupied space and the building as a whole.
D. There is no gross occupant load factor.

_______ 3. A means of egress consists of three separate and distinct parts. They are:
A. exit access, exit, exit discharge.
B. exit location, exit width, exit elevation.
C. hallways, stairways, doorways.
D. protected, unprotected, ventilated.

_______ 4. The length of the exit access establishes:
A. occupants allowed.
B. type of sprinklers required.
C. travel distance to an exit.
D. number of sprinklers required.

_______ 5. In most cases, the exit travel distance can be increased up to what percentage if the building is completely protected with an approved supervised automatic sprinkler system?
A. 10 percent
B. 25 percent
C. 50 percent
D. 75 percent

_______ 6. The width of an exit access should be at least sufficient for the number of persons it must accommodate:
A. The occupants within 30 feet of the exit
B. The occupants on that floor
C. Half of the number of allowable occupants
D. The occupant load of the room(s)

_______ 7. The specific placement of exits is a matter of design judgment, given the specifications of travel distance, allowable dead ends, common path of travel, and:
A. exit capacity.
B. elevation of structure.
C. landscape of building.
D. interior design.

_____ 8. If any exits discharge to the street floor, the following condition(s) must be satisfied:
- **A.** The exits must discharge to a free and unobstructed public way outside of the building.
- **B.** The street floor must be protected by automatic sprinklers.
- **C.** The street floor must be separated from any floors below by construction having a 2-hour fire resistance rating.
- **D.** All of the above.

_____ 9. The number of means of egress from any area is how many, unless specifically allowed by the *Life Safety Code*?
- **A.** 1
- **B.** At least 2
- **C.** More than 2
- **D.** 3

_____ 10. In reviewing plan sets for compliance with the travel distance limitations established for any occupancy, it is important to know:
- **A.** the type of exits.
- **B.** the type of flooring.
- **C.** the natural path of travel.
- **D.** the exterior landscape.

Fill in the Blank

Read each item carefully, and then complete the statement by filling in the missing word(s).

1. In reviewing plan sets for compliance with the travel distance limitations established for any occupancy, it is important to know the __________ __________ __________ __________ and the obstacles that will be present.
2. The __________ __________ is the number of people allowed through any point of the means of egress, such as doors, corridors, and stairs.
3. __________ __________ __________ is defined as the maximum number of people that are allowed through a specific egress point, such as a door, set of stairs, or a corridor.
4. When a door is marked as an exit, certain requirements in the __________ __________ __________ take effect such as identification, clear access, locking, and widths.
5. Egress doors in assembly and educational occupancies, such as schools or movie theaters, must be equipped with panic hardware when the __________ __________ exceeds 100.
6. With a __________ __________, space must be provided in the area of refuge for the people entering the refuge area in addition to the normal occupant load.
7. __________ __________ width must be based on calculated occupant loads.
8. __________ __________ __________ provide the highest protected type of stair enclosure recommended by the *Life Safety Code*.
9. A hallway, corridor, passage, tunnel, or underfloor or overhead passageway may be used as an __________ __________, providing it is separated and arranged according to the requirements for exits.
10. Fire escapes are, at best, a __________ __________ for standard interior or exterior stairs.

True/False

If you believe the statement to be more true than false, write the letter "T" in the space provided. If you believe the statement to be more false than true, write the letter "F."

_____ 1. The means of egress must be sized to accommodate all people occupying the area it serves.

_____ 2. Multiple exits are not required to provide alternative routes in case one exit becomes obstructed by fire and becomes unavailable.

_______ **3.** The exit access may be a corridor, aisle, balcony, gallery, room, porch, or roof.

_______ **4.** History shows that more than 25 percent of occupants will make an attempt to exit a building the same way they entered.

_______ **5.** The *Life Safety Code* requires that whenever multiple exits are required, they must be separated by a distance equivalent to half of the diagonal of the space being served.

_______ **6.** Horizontal exits cannot comprise more than 50 percent of the total required exit capacity, except in healthcare facilities.

_______ **7.** The best location for fire escapes is on exterior masonry walls without exposing windows, with access to fire escape balconies by exterior fire doors.

_______ **8.** An area of refuge cannot be any floor where an approved supervised sprinkler system is located.

_______ **9.** All required exits and access ways must be identified by readily visible signs when the exit or the way to reach it is not immediately visible to the occupants.

_______ **10.** Maintaining the means of egress in safe operating condition at all times is as important as the proper design of the egress system itself.

Vocabulary

Define the following terms using the space provided.

1. Area of refuge:

____

____

____

2. Exit discharge:

____

____

____

3. Horizontal exit:

____

____

____

4. Means of egress:

____

____

____

Short Answer

Complete this section with short written answers, using the space provided.

1. What items does NFPA 101 *Life Safety Code* address?

2. What is occupant load based on?

3. Identify and briefly describe the three separate and distinct parts of a means of egress.

4. How is travel distance to an exit established?

5. How can we calculate the egress capabilities of a door? How do we calculate the required width of stairs?

Fire Alarm

The following case scenario will give you an opportunity to explore the concerns associated with being a fire inspector. Read the scenario, and then answer each question in detail.

You are reviewing building plans for a new high-rise that will be built in your jurisdiction.

1. How will you determine the required stair width for the exit stairways?

2. How will you determine the gross and net occupant load for this high-rise?

Challenging Question

An existing city owned a one-story building that is being converted into an exercise room with equipment. This will be a standalone, single-use occupancy. The room dimensions will be 150′ × 75′ (45 × 22.9 m). Using Table 1 (pp. 120–121), calculate the occupant load for this structure.

Fire Detection Systems

The Workbook Activities

The following activities have been designed to help you. Your instructor may require you to complete some or all of these activities as a regular part of your fire inspector training program. You are encouraged to complete any activity that your instructor does not assign as a way to enhance your learning in the classroom.

Chapter Review

The following exercises provide an opportunity to refresh your knowledge of this chapter.

Matching

Match each of the terms in the left column to the appropriate definition in the right column.

	Term		Definition
______	**1.** Double-action pull-station	**A.**	A fire alarm system that sounds an alarm in the building and transmits a signal to the fire department via a public alarm box system.
______	**2.** Noncoded alarm	**B.**	An automatic or manually operated device in a fire alarm system that, when activated, causes the system to indicate an alarm condition.
______	**3.** Coded system	**C.**	A manual fire alarm activation device that takes a single step—such as moving a lever, toggle, or handle—to activate the alarm.
______	**4.** Single-action pull-station	**D.**	A sensing device that responds when its operating element is heated to a predetermined temperature.
______	**5.** Ionization smoke detector	**E.**	A manual fire alarm activation device that requires two steps to activate the alarm. The person must push in a flap, lift a cover, or break a piece of glass before activating the alarm.
______	**6.** Alarm initiation device	**F.**	An off-premises facility that monitors alarm systems and is responsible for notifying the fire department of an alarm. These facilities may be geographically located some distance from the protected building(s).
______	**7.** Protected premises fire alarm system	**G.**	A device containing a small amount of radioactive material that ionizes the air between two charged electrodes to sense the presence of smoke particles.
______	**8.** Auxiliary system	**H.**	An alarm system that provides no information at the alarm control panel indicating where the activated alarm is located.
______	**9.** Central station	**I.**	A fire alarm system design that divides a building or facility into zones and has audible notification devices that can be used to identify the area where an alarm originated.
______	**10.** Fixed-temperature heat detector	**J.**	A fire alarm system that sounds an alarm only in the building where it was activated. No signal is sent out of the building.

CHAPTER

8

Multiple Choice

Read each item carefully, and then select the best response.

_____ 1. Either an automatic or manually operated device that, when activated, causes an alarm system to indicate an alarm:
   A. Alarm initiation device
   B. Protected premises fire alarm system
   C. Alarm notification device
   D. Alarm matrix

_____ 2. A chart showing what will happen with the fire alarm and detection system when an initiating device is activated is called:
   A. alarm initiation device.
   B. protected premises fire alarm system.
   C. alarm notification device.
   D. alarm matrix.

_____ 3. This device serves as the brain of the system, linking the activation devices to the notification devices:
   A. Alarm initiation device
   B. Fire alarm control panel
   C. Alarm notification device
   D. Alarm matrix

_____ 4. This panel enables fire fighters to ascertain the type and location of the activated alarm device as they enter the building, eliminating the need for fire fighters to hunt down the control panel to determine the problem:
   A. Alarm initiation device
   B. Fire alarm control panel
   C. Remote annunciator
   D. Alarm matrix

_____ 5. These detectors work on the principle that burning materials release many different products of combustion, including electrically charged microscopic particles:
   A. Photoelectric smoke detectors
   B. Ionization smoke detectors
   C. Hard-wired detectors
   D. Battery-operated detectors

_____ 6. These detectors alarm when visible particles of smoke pass through a light beam, interfering with the light beam and activating the alarm:
   A. Photoelectric smoke detectors
   B. Ionization smoke detectors
   C. Hard-wired detectors
   D. Battery-operated detectors

_____ 7. These devices are designed so that building occupants can activate the fire alarm system on their own if they discover a fire in the building.
- A. Smoke detectors
- B. Protected premises fire alarm systems
- C. Automatic alarm systems
- D. Manual initiation devices

_____ 8. These devices are designed to function without human intervention and activate the alarm system when they detect evidence of smoke or fire.
- A. Smoke detectors
- B. Protected premises fire alarm systems
- C. Automatic initiation devices
- D. Manual initiation devices

_____ 9. This device is a type of photoelectric smoke detector used to protect large spans such as churches, auditoriums, airport terminals, and indoor sports arenas.
- A. Heat detector
- B. Beam detector
- C. Rate of rise detector
- D. Line detector

_____ 10. These devices are designed to operate at a preset temperature.
- A. Fixed-temperature heat detectors
- B. Beam detector
- C. Rate of rise detector
- D. Line detector

_____ 11. These devices are individual units that can be spaced throughout an occupancy, so that each detector covers a specific floor area.
- A. Spot detectors
- B. Flame detector
- C. Rate of rise detector
- D. Line detector

_____ 12. These detectors use wire or tubing strung along the ceiling of large open areas to detect an increase in heat.
- A. Spot detectors
- B. Beam detector
- C. Rate of rise detector
- D. Line detector

_____ 13. These detectors are specialized devices that detect the electromagnetic light waves produced by a flame.
- A. Spot detectors
- B. Flame detector
- C. Rate of rise detector
- D. Line detector

_____ 14. These detectors are calibrated to detect the presence of a specific gas that is created by combustion or used in the facility.
- A. Spot detectors
- B. Flame detector
- C. Gas detector
- D. Line detector

Fill in the Blank

Read each item carefully, then complete the statement by filling in the missing word(s).

1. ____________ ____________ detectors continuously capture air samples and measure the concentrations of specific gases or products of combustion.

2. When air-sampling detectors are installed in the return air ducts of large buildings, they are known as ____________ ____________.

3. Automatic sprinkler systems are usually connected to the fire alarm system through a water flow paddle and will ________________ ________________ ________________ if a water flow occurs.

4. Model building and fire prevention codes have adopted a standardized audio pattern, called the ________________ ________________, that must be produced by any audio device used as a fire alarm.

5. Many new fire alarm systems incorporate ________________ ________________ devices such as high-intensity strobe lights as well as audio devices.

6. In a ________________ ________________, the alarm control panel indicates where in the building the alarm was activated.

7. In a ________________ ________________, the zone is identified not only at the alarm control panel but also throughout the building using the audio notification device.

8. In a ________________ alarm system, the control panel has no information indicating where in the building the fire alarm was activated.

9. ________________ ________________ alarm system buildings are divided into multiple zones, often by floor or by wing. The alarm control panel indicates in which zone the activated device is located.

10. Hospitals often use a ________________ ________________ alarm system because it is not possible to evacuate all staff and patients for every fire alarm.

True/False

If you believe the statement to be more true than false, write the letter "T" in the space provided. If you believe the statement to be more false than true, write the letter "F."

______ 1. In a master-coded alarm system, the audible notification devices are used for fire alarms only.

______ 2. The fire department should always be notified when a fire alarm system is activated.

______ 3. A protected premises fire alarm system sounds only in the building to notify the occupants, not the fire department.

______ 4. A remote supervising station system sends a signal directly to the occupants via a telephone line or a radio signal.

______ 5. Auxiliary systems would only be seen in large older cities because most cities have phased out public fire alarm boxes.

______ 6. In a proprietary supervising system, the building's fire alarms are connected directly to the fire department. Proprietary systems are often installed in small single-occupancy structures.

______ 7. A central station is a third-party off-site monitoring facility that monitors multiple alarm systems. Individual building owners contract and pay the central station to monitor their facilities.

______ 8. There should be a monthly visual inspection of alarm systems looking for items such as visibility of pull-stations, access to the pull-stations, smoke detectors with the caps still in place, damage to any devices, and anything else that might prohibit smoke or heat detectors from operating such as painted heat detectors and bags on smoke detectors.

______ 9. Batteries in fire detection systems should be replaced about every 7 years.

______ 10. As a fire inspector, you should conduct all tests on fire detection systems in your jurisdiction.

Labeling

Label the following detectors with the correct terms.

1. ______________________________

2. ______________________________

3. ______________________________

4. ______________________________

5. ______________________________

Vocabulary

Define the following terms using the space provided.

1. Auxiliary system:

____

____

____

2. Bimetallic strip:

____

____

____

3. Ionization smoke detector:

____

____

____

4. T tapping:

5. Temporal-3 pattern:

Short Answer

Complete this section with short written answers using the space provided.

1. Identify the two different types of smoke detectors and briefly describe their features and applications.

2. Identify and briefly describe the three most common causes of problems with smoke detectors.

3. Identify and briefly describe the four categories of alarm annunciation systems.

4. Identify and briefly describe the five categories of fire department notification systems.

Fire Alarm

The following case scenario will give you an opportunity to explore the concerns associated with being a fire inspector. Read the scenario and then answer each question in detail.

You are reviewing the annual inspection paperwork for an alarm system that is in service at your local shopping mall. There were several deficiencies noted during the annual inspection.

1. How should you handle the noted deficiencies?

You have been assigned to do a preoccupancy inspection of a newly installed fire detection system in an office complex.

2. Briefly describe your concerns with a new system and how you will conduct your preoccupancy inspection.

Fire Flow and Fire Suppression Systems

The Workbook Activities

The following activities have been designed to help you. Your instructor may require you to complete some or all of these activities as a regular part of your fire inspector training program. You are encouraged to complete any activity that your instructor does not assign as a way to enhance your learning in the classroom.

Chapter Review

The following exercises provide an opportunity to refresh your knowledge of this chapter.

Matching

Match each of the terms in the left column to the appropriate definition in the right column.

_____	**1.** Water main	**A.**	A valve that signals an alarm when a sprinkler head is activated and prevents nuisance alarms caused by pressure variations.
_____	**2.** Dry-pipe valve	**B.**	A valve that allows flow in one direction only.
_____	**3.** Wall post indicator valve (WPIV)	**C.**	A sprinkler control valve with a valve stem that moves in and out as the valve is opened or closed.
_____	**4.** Outside stem and yoke (OS&Y) valve	**D.**	A type of indicating valve that moves a piece of metal 90° within the pipe and shows if the water supply is open or closed.
_____	**5.** Alarm valve	**E.**	A valve assembly designed to release water into a sprinkler system when an external initiation device is activated.
_____	**6.** Shut-off valve	**F.**	A sprinkler control valve with an indicator that reads either open or shut depending on its position.
_____	**7.** Post indicator valve (PIV)	**G.**	A sprinkler control valve that is mounted on the outside wall of a building. The position of the indicator tells whether the valve is open or shut.
_____	**8.** Deluge valve	**H.**	The generic term for any underground water pipe.
_____	**9.** Butterfly valve	**I.**	Any valve that can be used to shut down water flow to a water user or system.
_____	**10.** Check valve	**J.**	The valve assembly on a dry sprinkler system that prevents water from entering the system until the air pressure is released.

Multiple Choice

Read each item carefully, and then select the best response.

_____ **1.** The evaluation of a fire suppression system begins in the:

- **A.** plan review stage.
- **B.** construction stage.
- **C.** planning stage.
- **D.** approval stage.

CHAPTER 9

______ 2. This type of water system provides water to fire hydrants under pressure:
- A. Municipal
- B. Static
- C. Reservoir
- D. Riser

______ 3. Most water distribution systems use a combination of these type systems:
- A. Dry and wet hydrant
- B. Small and large delivery tubes
- C. Direct pumping systems and gravity systems
- D. Risers and standpipes

______ 4. The underground water mains that deliver water to the end users come in several different sizes. The largest mains are called:
- A. primary feeders.
- B. secondary feeders.
- C. distributors.
- D. standpipe.

______ 5. These are used in locations where temperatures may drop below freezing:
- A. Wet barrel hydrants
- B. Dry barrel hydrants
- C. WPIV valves
- D. PIV valves

______ 6. The pressure in a water supply system when the water is not moving is called:
- A. residual pressure.
- B. flow pressure.
- C. back pressure.
- D. static pressure.

______ 7. The quantity of water flowing through an opening during a hydrant test is called:
- A. residual pressure.
- B. flow pressure.
- C. back pressure.
- D. static pressure.

______ 8. This tool is used to measure flow pressure in psi (or kilopascals) and to calculate the flow in gallons (or liters) per minute:
- A. Discharge valve
- B. Fire department connection
- C. Pitot gauge
- D. Ralph tool

______ 9. This occupancy hazard class is subdivided into two groups and may produce severe fires:
- A. Light-hazard class
- B. Ordinary hazard class
- C. Extra hazard occupancy class
- D. Special occupancy conditions

_____ 10. This is a necessary part of an automatic sprinkler system to ensure that the presence of the primary water supply is maintained:
- A. Fire department connection
- B. Post-indicator valve
- C. Outside stem and yoke valve
- D. Main valve

_____ 11. These are vertical sections of pipe that connect the underground supply to the rest of the sprinkler piping in the system:
- A. Deluge
- B. Cross main
- C. Riser
- D. Distributer

_____ 12. This is a mechanical alarm notification device that is powered by water moving through the sprinkler system:
- A. Clapper
- B. Butterfly
- C. Water horn
- D. Water motor gong

_____ 13. This type sprinkler system is the most common and least expensive type of automatic sprinkler system. When a sprinkler head activates, water is immediately discharged onto the fire:
- A. Wet-pipe sprinkler system
- B. Dry-pipe system
- C. Preaction systems
- D. Combined dry-pipe and preaction systems

_____ 14. In this type sprinkler system, the pipes are filled with pressurized air or nitrogen instead of water. A valve keeps water from entering the pipes until the air or nitrogen pressure drops below a threshold pressure:
- A. Wet-pipe sprinkler system
- B. Dry-pipe system
- C. Preaction systems
- D. Combined dry-pipe and preaction systems

_____ 15. This type sprinkler system is also known as an interlock system, and it is similar to a dry sprinkler system with one key difference: In this system, a secondary device—such as a smoke detector or a manual-pull alarm—must be activated before water is released into the sprinkler piping:
- A. Wet-pipe sprinkler system
- B. Dry-pipe system
- C. Preaction systems
- D. Combined dry-pipe and preaction systems

Fill in the Blank

Read each item carefully, and then complete the statement by filling in the missing word(s).

1. Large dry-pipe sprinkler systems can take several minutes to empty out the air and refill the pipes with water when a head is activated. To compensate for this problem, two additional devices are used: _______________ and _______________.
2. A _______________ _______________ uses a deluge valve instead of a dry-pipe valve.
3. A _______________ _______________ _______________ is a type of dry sprinkler system in which water flows from all the sprinkler heads as soon as the system is activated.
4. Specialized extinguishing systems are often used in areas where _______________ would not be an acceptable extinguishing agent.
5. Dry-chemical extinguishing systems use the same types of finely powdered agents as _______________ _______________ fire extinguishers.

6. Only wet agents such as class K or B:C-rated dry chemical extinguishing agents should be used where ____________ ____________ extinguishing systems are installed.

7. Clean agent extinguishing systems are often installed in areas where ____________ ____________ ____________ are used or where valuable documents are stored.

8. Until the 1990s, ____________ ____________ was the agent of choice for protecting areas such as computer rooms, telecommunications rooms, and other sensitive areas.

9. ____________ ____________ ____________ usually have the same series of pre-alarms and abort buttons as are found in Halon 1301 systems.

10. To perform a dry-pipe system air test, air pressure is pumped up to ____________ ____________ in the piping for ____________ ____________.

True/False

If you believe the statement to be more true than false, write the letter "T" in the space provided. If you believe the statement to be more false than true, write the letter "F."

_____ 1. The upright sprinkler head are designed to be installed so that the water spray is directed downward against the deflector.

_____ 2. A standpipe system consists of a network of inlets, pipes, and outlets for fire hoses that are built into a structure to provide water for firefighting purposes.

_____ 3. Pendent sprinkler heads are designed to be installed so that the water stream is directed downward against the deflector.

_____ 4. A hydrostatic sprinkler system test is a two-hour test conducted to make certain that the system components in a sprinkler system will not leak.

_____ 5. Ornamental sprinklers are sprinklers in which all or part of the body, including the shank thread, is mounted above the lower plane of the ceiling.

_____ 6. A Class I standpipe is designed for use by fire department personnel only, and each outlet has a 2½" (64 mm) male coupling and a valve to open the water supply after the attack line is connected.

_____ 7. Intermediate-level sprinklers are sprinklers that have been specifically designed and listed for use in residential occupancies.

_____ 8. Both standpipe systems and sprinkler systems are supplied with water in essentially the same way.

_____ 9. Fusible-link sprinkler heads use a glass bulb filled with glycerin or alcohol to hold the cap in place.

_____ 10. A class III standpipe has the features of both class I and class II standpipes in a single system. This kind of system has 2½" (64 mm) outlets for fire department use as well as smaller outlets with attached hoses for occupant use.

Labeling

Label the following sprinklers with the correct terms.

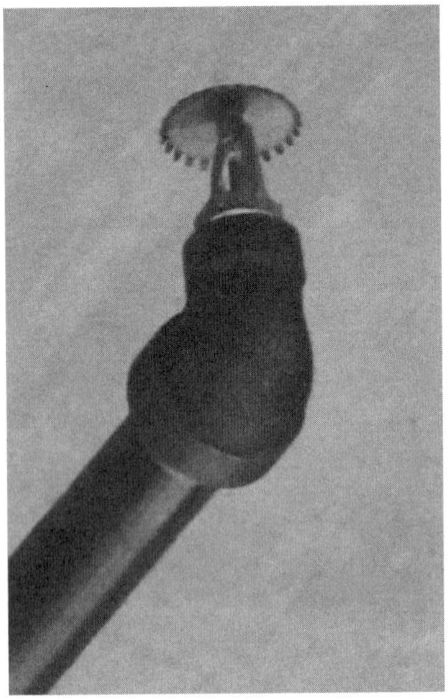

1. ______________________________

2. ______________________________

3. ______________________________

4. ______________________________

5. ______________________________

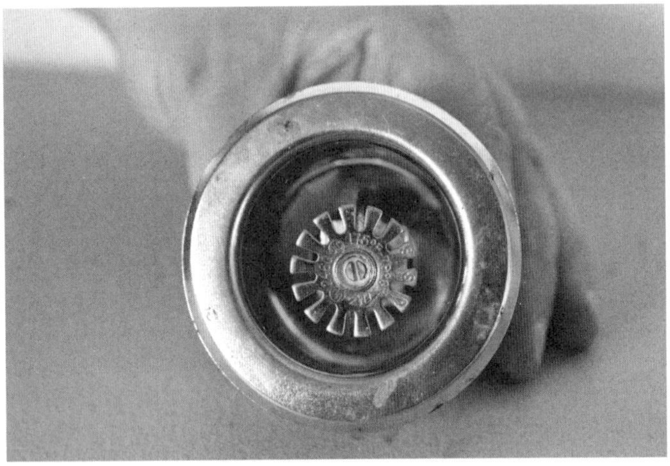

6. ______________________________

7. ______________________________

8. ______________________________

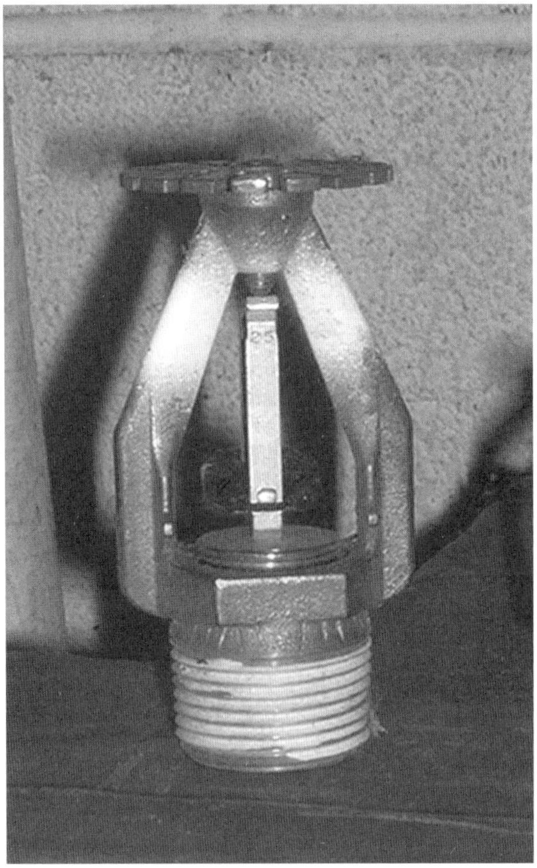

9. ______________________________

10. ______________________________

11. ______________________________

12. ______________________________

Vocabulary

Define the following terms using the space provided.

1. Accelerator:

____

____

____

2. Distributors:

____

____

____

3. Exhauster:

____

____

____

4. Halon 1301:

____

____

____

5. Riser:

____

____

____

Short Answer

Complete this section with short written answers using the space provided.

1. Briefly explain the purpose and usage of elevated water storage towers.

2. NFPA 24, *Standard for the Installation of Private Fire Service Mains and Their Appurtenances*, recommends that fire hydrants be color coded to indicate the water flow available from each hydrant at 20 psi. What are the colors and flow?

3. Briefly describe the differences between static pressure, elevation pressure, residual pressure, and flow pressure.

4. What is a Pitot gauge, and how is it used?

5. What is the easiest way to determine the flow of a fire hydrant?

Fire Alarm

The following case scenario will give you an opportunity to explore the concerns associated with being a fire inspector. Read the scenario, and then answer each question in detail.

You are reviewing building plans for a new high-rise that will be built in your jurisdiction.

1. How will you determine the specific requirements for the spacing of sprinkler heads, the sprinkler head discharge densities, and water supply requirements?

2. How will you determine the proper suppression systems to be used at an outside gasoline fill station and the kitchen area of this new high-rise?

____

____

____

____

____

____

____

____

____

____

Portable Fire Extinguishers

The Workbook Activities

The following activities have been designed to help you. Your instructor may require you to complete some or all of these activities as a regular part of your fire inspector training program. You are encouraged to complete any activity that your instructor does not assign as a way to enhance your learning in the classroom.

Chapter Review

The following exercises provide an opportunity to refresh your knowledge of this chapter.

Matching

Match each of the terms in the left column to the appropriate definition in the right column.

_____	**1.** Underwriters Laboratories, Inc. (UL)	**A.**	A water-based fire extinguisher that uses an alkali metal salt as a freezing point depressant.
_____	**2.** Aqueous film-forming foam (AFFF)	**B.**	An extinguishing agent for class K fires. It commonly consists of solutions of water and potassium acetate, potassium carbonate, potassium citrate, or any combination thereof.
_____	**3.** Dry-powder extinguishing agent	**C.**	A fire extinguisher that uses carbon dioxide gas as the extinguishing agent.
_____	**4.** Film-forming fluoroprotein (FFFP) foam	**D.**	An extinguisher that uses a mixture of finely divided solid particles to extinguish fires. The agent is usually based on sodium bicarbonate, potassium bicarbonate, or ammonium phosphate, with additives included to provide resistance to packing and moisture absorption and to promote proper flow characteristics.
_____	**5.** Multipurpose dry chemical fire extinguisher	**E.**	The U.S. organization that tests and certifies that fire extinguishers (among many other products) meet established standards. The Canadian equivalent is Underwriters Laboratories of Canada (ULC).
_____	**6.** Loaded-stream fire extinguisher	**F.**	A water-based extinguishing agent used on class B fires that forms a foam layer over the liquid and stops the production of flammable vapors.
_____	**7.** Halogenated extinguishing agent	**G.**	A water-based extinguishing agent used on class B fires that forms a foam layer over the liquid and stops the production of flammable vapors.
_____	**8.** Carbon dioxide (CO_2) fire extinguisher	**H.**	An extinguishing agent used in putting out class D fires. The common dry powder extinguishing agents include powders based on sodium chloride and graphite.
_____	**9.** Dry chemical fire extinguisher	**I.**	A fire extinguisher rated to fight class A, B, and C fires.
_____	**10.** Wet chemical extinguishing agent	**J.**	A liquefied gas extinguishing agent that puts out fires by chemically interrupting the combustion reaction between the fuel and oxygen.

CHAPTER 10

Multiple Choice

Read each item carefully, and then select the best response.

_____ **1.** Portable fire extinguishers have one primary use, which is:
- **A.** to extinguish fully involved fires.
- **B.** to extinguish incipient fires.
- **C.** to extinguish class B fires.
- **D.** to extinguish class K fires.

_____ **2.** This class fire involves ordinary combustibles such as wood, paper, cloth, rubber, household rubbish, and some plastics.
- **A.** Class A
- **B.** Class B
- **C.** Class C
- **D.** Class K

_____ **3.** This class fire involves combustible cooking oils and fats.
- **A.** Class A
- **B.** Class B
- **C.** Class C
- **D.** Class K

_____ **4.** A fire extinguisher that carries a 2-A:10-B:C rating can be used on which class fire(s)?
- **A.** Class A
- **B.** Class B
- **C.** Class C
- **D.** Class A, class B, and class C

_____ **5.** A 10-B rating indicates that a nonexpert user should be able to extinguish a fire in a pan of flammable liquid that is how many square feet in surface area?
- **A.** 1 ft^2
- **B.** 10 ft^2
- **C.** 100 ft^2
- **D.** 1000 ft^2

_____ **6.** Numbers are used to rate an extinguisher's effectiveness only for what class fires?
- **A.** Class A and class B fires
- **B.** Class C and class D fires
- **C.** Class C, class D, and class K fires
- **D.** Class A, class B, and class K fires

_____ **7.** In areas that are subject to below-freezing temperatures, what type fire extinguishers can be used to counteract freezing of contents?
- **A.** Dry chemical fire extinguishers
- **B.** Loaded-stream fire extinguishers
- **C.** High-pressure water extinguishers
- **D.** Inversion-type fire extinguishers

_____ 8. What is the only dry chemical extinguishing agent that is rated as suitable for use on class A fires?
A. Carbon monoxide
B. Carbon dioxide
C. Ammonium phosphate
D. Dry chemical

_____ 9. Carbon dioxide is rated for what classes of fire?
A. Class A and C fires only
B. Class B and K fires only
C. Class B and C fires only
D. All classes of fires

_____ 10. What is the only type of extinguisher to qualify under the new class K rating requirements?
A. Carbon dioxide
B. Aqueous film-forming foam
C. Halogenated extinguishing agents
D. Wet chemical fire extinguishers

_____ 11. Wet agents convert the fatty acids in cooking oils or fats to a soap or foam in a process known as:
A. saponification.
B. acetation.
C. goopification.
D. distillation.

_____ 12. Dry-powder extinguishing agents are chemical compounds used to extinguish fires involving what type of fuel?
A. Ordinary combustibles
B. Flammable or combustible liquids
C. Energized electrical equipment
D. Combustible metals

Fill in the Blank

Read each item carefully, and then complete the statement by filling in the missing word(s).

1. Fire extinguishers can be used to control fires where _______________ extinguishing methods are not recommended.

2. It is essential to match the appropriate type of extinguisher to the _______________ _______________ _______________.

3. Portable fire extinguishers are classified and rated based on their _______________ and _______________.

4. The number on the class A and class B fire extinguishers indicates the _______________ _______________ of the fire extinguisher in the hands of a nonexpert user.

5. Fire extinguishers are rated for their ability to control a specific type of fire as well as for the extinguishing agent's ability to _______________ _______________.

6. Under the pictograph labeling system, the presence of an icon indicates that the extinguisher has been rated for that class of fire. A missing icon indicates that the extinguisher has _______________ _______________ _______________ for that class of fire.

7. Two key factors must be considered when determining which type of extinguisher should be placed in each area: the class of fire that is likely to occur and the _______________ _______________ of an incipient fire.

8. Wetting agents reduce the _______________ _______________ of the water, allowing it to penetrate more effectively into many fuels, such as baled cotton or fibrous materials.

9. Dry chemical extinguishing agents work to interrupt the ________________ ________________ ________________ that occur as part of the combustion process.

10. Some class B foam extinguishing agents are approved for use on ________________ ________________, that is, water-soluble flammable liquids such as alcohols, acetone, esters, and ketones.

True/False

If you believe the statement to be more true than false, write the letter "T" in the space provided. If you believe the statement to be more false than true, write the letter "F."

_____ 1. The occupancy use category does not necessarily determine the building's hazard classification.

_____ 2. Class A foams are very effective on class B fires.

_____ 3. Halogenated extinguishing agents release a mist of vapor and liquid droplets that disrupts the molecular chain reactions within the combustion process, thereby extinguishing the fire.

_____ 4. When halogenated extinguishing agents are applied to burning vegetable oils, they create a thick blanket of foam that quickly smothers the fire and prevents it from reigniting while the hot oil cools.

_____ 5. Dry chemical extinguishing agents are chemical compounds used to extinguish fires involving combustible metals (class D fires).

_____ 6. Wet chemical fire extinguishers are the only type of extinguisher to qualify under the new class K rating requirements.

_____ 7. A blanket of carbon dioxide over the surface of a liquid fuel can disrupt the fuel's ability to vaporize.

_____ 8. Per pound, halogenated extinguishing agents are approximately half as effective at extinguishing fires as carbon dioxide.

_____ 9. Water is an excellent extinguishing agent for class D fires.

_____ 10. Ammonium phosphate is the only dry chemical extinguishing agent rated as suitable for use on class A fires.

Vocabulary

Define the following terms using the space provided.

1. Aqueous film-forming foam (AFFF):

____

____

____

2. Clean agent:

____

____

____

3. Polar solvent:

____

____

____

4. Saponification:

Short Answer

Complete this section with short written answers using the space provided.

1. List the five types of fires by class.

2. List the traditional lettering and labeling system used to identify portable extinguishers.

3. Describe how the hazard classification for each area of a structure should be evaluated when determining the type and size extinguishers to be placed in these areas.

4. Identify the seven basic types of extinguishing agents that portable fire extinguishers use.

5. What are the limitations and disadvantages of carbon dioxide extinguishers?

Fire Alarm

The following case scenario will give you an opportunity to explore the concerns associated with being a fire inspector. Read the scenario and then answer each question in detail.

You have been assigned the task of evaluating the state of readiness of extinguishers in an industrial location in your jurisdiction. As a newly trained and certified inspector, you have never inspected extinguishers. This will be your first extinguisher inspection.

1. Develop a list of items you should inspect. Include damage assessment and any other items that should be brought to the attention of the industrial complex safety officer.

2. Identify common problems and deficiencies normally uncovered during inspections of extinguishers. How should these and other noted problems be handled?

Electrical and HVAC Hazards

The Workbook Activities

The following activities have been designed to help you. Your instructor may require you to complete some or all of these activities as a regular part of your fire inspector training program. You are encouraged to complete any activity that your instructor does not assign as a way to enhance your learning in the classroom.

Chapter Review

The following exercises provide an opportunity to refresh your knowledge of this chapter.

Matching

Match each of the terms in the left column to the appropriate definition in the right column.

______	1. HVAC	A.	A closed vessel in which water is heated, steam is generated, steam is superheated, or any combination thereof by the application of heat from combustible fuels in a self-contained or attached furnace.
______	2. Air handling unit	B.	The subdivision of a building into relatively small areas so that fire or smoke can be confined to the room or section in which it originates.
______	3. Equipment ground-fault protective device (EGFPD)	C.	A device designed to open and close a circuit by nonautomatic means and to open the circuit automatically on a predetermined overcurrent without damage to itself when properly applied within its rating.
______	4. Circuit breakers	D.	Abbreviation for heating, ventilation, and air conditioning systems.
______	5. Flue	E.	A unit installed for the purpose of processing the treatment of air so as to control simultaneously its temperature, humidity, and cleanliness to meet the requirements of the conditioned space it serves.
______	6. Immediately Dangerous to Life or Health (IDLH)	F.	A device intended for protection of personnel that functions to deenergize a circuit or portion thereof within an established period of time when a fault current-to-ground exceeds some predetermined value that is less than that required to operate the overcurrent protective device of that supply circuit.
______	7. Plenum system	G.	The general term for a passage through which flue gases are conveyed from the combustion chamber to the outer air.
______	8. Boiler	H.	An HVAC system that uses a compartment or chamber to which one or more air ducts are connected and that forms part of the air distribution system.
______	9. Ground-fault circuit interrupter (GFCI)	I.	A device intended to provide protection of equipment from damage from line-to-ground fault currents by operating to cause a disconnecting means to open all ungrounded conductors of the faulted circuit.
______	10. Compartmentation	J.	Any atmosphere that poses an immediate hazard to life or produces immediate irreversible debilitating effects on health.

CHAPTER 11

Multiple Choice

Read each item carefully, and then select the best response.

______ **1.** A device intended to provide protection from the effects of arc faults by recognizing characteristics unique to arcing and by functioning to deenergize the circuit when an arc fault is detected is a (an):
- **A.** ground-fault circuit interrupters (GFCIs).
- **B.** arc-fault circuit interrupter (AFCI).
- **C.** equipment ground-fault protective devices (EGFPDs).
- **D.** circuit breaker.

______ **2.** The volume of electrical flow is measured in what unit?
- **A.** Watts
- **B.** Volts
- **C.** Arcs
- **D.** Ampere (amps or amperage)

______ **3.** The permanent joining of metallic parts to form an electrically conductive path that will ensure electrical continuity and the capacity to conduct any current likely to be imposed is:
- **A.** bonding.
- **B.** grounding.
- **C.** conduit.
- **D.** damper.

______ **4.** This helps to eliminate electrical shock hazard by providing a path for stray currents and any accumulation of static electricity to flow.
- **A.** Bonding
- **B.** Grounding
- **C.** Conduit
- **D.** Damper

______ **5.** Some circuit breakers may have this device that allows them to be operated from remote locations.
- **A.** GFCI
- **B.** AFCI
- **C.** Shunt trip
- **D.** Cartridge fuse

______ **6.** These are more common in industrial facilities and as the main transformer on incoming services where large current loads are present due to their more efficient operation.
- **A.** Dry-type transformers
- **B.** Transfer switches
- **C.** Fluid-filled transformers
- **D.** Emergency generators

______ **7.** Outdoor transformers should be located in such a way that leaking fluids will do what?
- **A.** Be contained in an aboveground diked area
- **B.** Be contained below the transformer
- **C.** Be cycled back into the transformer
- **D.** Drain away from buildings and prevented from entering environmentally sensitive areas

_____ **8.** Which of the following are *not* components of a structural electrical system?
A. Conduit
B. Raceway
C. Vestibule
D. Junction box

_____ **9.** During a fire inspection, what should circuit breaker boxes be checked for?
A. Obvious deterioration, dirt
B. Moisture, tracking
C. Poor maintenance
D. All of the above

_____ **10.** Which professional individuals should not be allowed to test for static charge in electrically hazardous locations?
A. Industrial hygienists
B. Safety engineers
C. Electrical engineers
D. Fire inspectors

_____ **11.** What is the name of an object that, through conduction, draws heat away from a heat-producing object?
A. Heat sync
B. Junction box
C. Raceway
D. Grounding rod

_____ **12.** Which of the following is not a possible component of exhaust systems?
A. Dust collection system
B. Chemical treatment system
C. Chiller plant system
D. Damper

Fill in the Blank

Read each item carefully, and then complete the statement by filling in the missing word(s).

1. _______________ _______________ verify compliance with adopted electrical code when buildings are initially constructed, expanded, altered, or renovated.

2. The force required to move or conduct electricity is measured in units called _______________.

3. The total electrical power available for use is a combination of the volts and the amps, the _______________.

4. A _______________ is an overcurrent protective device with a circuit-opening fusible part that is heated and severed by the passage of overcurrent through it.

5. The _______________ _______________ operates based on a variety of principles, but all have the same result: A switch is opened to stop the flow of electricity through the circuit.

6. Ground-fault circuit interrupters (GFCIs) sense when the current passes to _______________ through any path other than the proper path.

7. A metallic underground _______________ _______________ _______________ must be used as the grounding electrode where it is available and where the buried portion of the pipe is more than 10' (3 m) long.

8. Emergency generator storage tanks must be inspected for spill protection, venting requirements, and fixed-fire protection if the volume of the storage tank exceeds the limits imposed by the _______________ _______________ _______________.

9. Direct connection of the emergency generator to the building's electrical system should only be accomplished through a ____________ ____________ installed in accordance with NFPA 70, *National Electrical Code*.

10. Electric cables should be protected from ____________ ____________ where they pass through walls or floors.

True/False

If you believe the statement to be more true than false, write the letter "T" in the space provided. If you believe the statement to be more false than true, write the letter "F."

_____ 1. Extension cords should be used only to connect permanent portable equipment, not as a substitute for temporary wiring.

_____ 2. Raceways generate steam, hot water, or hot air by consuming electricity or burning natural gas, coal, or fossil fuels such as kerosene.

_____ 3. An outlet is any set of contacts that interrupts or controls current flow through an electrical circuit.

_____ 4. Hydronic heating and cooling systems circulate hot or chilled water through plastic piping embedded in a gypsum/cement floor layer or imbedded in a concrete floor slab.

_____ 5. Activation of fire dampers is accomplished by the use of a fusible link or a thermal sensor in power-operated dampers.

_____ 6. Electrically hazardous areas are those in which flammable liquids, gases, combustible dusts, or readily ignitable fibers are present in sufficient quantities to represent a fire or explosion hazard.

_____ 7. Ducted systems are used to distribute conditioned air from air-handling units or indirectly heated and cooled air from furnaces, air-conditioning coils, and other heat exchangers.

_____ 8. Any indication of exhaust stack damage, such as broken or separated flue piping, damaged masonry, or a leaning exhaust structure, should be further investigated by a qualified mechanical or fire inspector.

_____ 9. Boxes and cabinets have prepunched circular holes called knockouts that can be removed to allow the secure connection of conduit or raceways.

_____ 10. Methods to bring the hazard of static electricity under reasonable control are humidification, bonding, grounding, ionization, conductive floors, or a combination of these methods.

Vocabulary

Define the following terms using the space provided.

1. Amps:

____

____

____

2. Conduits:

____

____

____

3. Gravity vents:

4. Plenum system:

5. Raceway:

Short Answer

Complete this section with short written answers using the space provided.

1. Identify four warning signs of possible electrical problems.

2. Identify and briefly describe two potential electrical hazards.

3. Identify and briefly describe the two most commonly used overcurrent devices.

4. Identify some common problems found when inspecting electrical wiring.

5. Identify and describe four smoke management systems used to work with a buildings construction to aid in the movement of smoke.

Fire Alarm

The following case scenario will give you an opportunity to explore the concerns associated with being a fire inspector. Read the scenario, and then answer each question in detail.

You are doing an initial inspection of an industrial site that has just been annexed into your jurisdiction. The site has several old structures with old HVAC systems, as well as old-style transformers and generators. The site also has several new buildings, complete with new HVAC systems, electrical systems, and fire suppression systems.

1. Identify the fire and safety concerns you have with both the older style transformers you are inspecting as well as the new transformers you see on site.

____

____

____

____

____

____

____

____

____

____

2. One of the more recently built multistory structures on site has a smoke management system installed in the stairwells. How will you test these systems, and where will you find reference materials to help you?

Ensuring Proper Storage and Handling Practices

The Workbook Activities

The following activities have been designed to help you. Your instructor may require you to complete some or all of these activities as a regular part of your fire inspector training program. You are encouraged to complete any activity that your instructor does not assign as a way to enhance your learning in the classroom.

Chapter Review

The following exercises provide an opportunity to refresh your knowledge of this chapter.

Matching

Match each of the terms in the left column to the appropriate definition in the right column.

	Term	Definition
______	**1.** Shipping papers	**A.** A refrigerated liquid gas having a boiling point below –130°F (–90°C) at atmospheric pressure.
______	**2.** Liquefied gas	**B.** Information on a pesticide label that indicates the relative toxicity of the material.
______	**3.** Intermodal tanks	**C.** A gas, other than in solution, that in a packaging under the charged pressure exists both as a liquid and a gas at a temperature of 20°C (68°F).
______	**4.** Bills of lading	**D.** Tanks designed to carry dry bulk goods such as powders, pellets, fertilizers, or grain; they are generally V-shaped with rounded sides that funnel toward the bottom.
______	**5.** Combustible liquids	**E.** Bulk containers that can be shipped by all modes of transportation—air, sea, or land.
______	**6.** Fuel gases	**F.** Any gas used as a fuel source, including natural gas, manufactured gas, sludge gas, liquefied petroleum gas-air mixtures, liquefied petroleum gas in the vapor phase, and mixtures of these gases.
______	**7.** Signal words	**G.** An explosion that occurs when a tank containing a volatile liquid is heated.
______	**8.** Cryogenic gas	**H.** Any liquid that has a closed-cup flash point at or above 100°F (37.8°C).
______	**9.** Dry bulk cargo tanks	**I.** A shipping order, bill of lading, manifest, or other shipping document serving a similar purpose and usually including the names and addresses of both the shipper and the receiver as well as a list of shipped materials with quantity and weight.
______	**10.** Boiling liquid expanding vapor explosion (BLEVE)	**J.** Shipping papers for roads and highways.

CHAPTER 12

Multiple Choice

Read each item carefully, and then select the best response.

_____ 1. Any liquids having a flash point below 100°F (37.8°C) and having a vapor pressure not exceeding 40 psi (276 kPa) (absolute) at 100°F (37.8°C) are:
- A. combustible liquids.
- B. flammable liquids.
- C. cryogenic liquids.
- D. compressed liquids.

_____ 2. What are shipping papers for roads and highways called?
- A. Freight bills
- B. Shipping bills
- C. Consists
- D. Waybills

_____ 3. This form provides basic information about the chemical makeup of a substance, the potential hazards it presents, appropriate first aid in the event of an exposure, and other pertinent data for safe handling of the material.
- A. SDS
- B. Bills of lading
- C. Chem bills
- D. Waybills

_____ 4. This form may have additional information about a hazardous substance such as its packaging group designation.
- A. SDS
- B. Bills of lading
- C. Chem bills
- D. Waybills

_____ 5. Flammable liquid storage tanks can be installed where?
- A. Above ground
- B. Underground
- C. Inside buildings
- D. All of the above

_____ 6. Within the military marking system, materials that are considered mass detonation hazards are classified in which division?
- A. Division 1
- B. Division 2
- C. Division 3
- D. Division 4

_____ 7. In the military system chemical hazards such as sarin or mustard gas are depicted by what colors?
- A. Red
- B. Yellow
- C. White
- D. Green

_____ 8. When looking at the NFPA 704 marking system, numbers in the blue diamond indicates hazard for what hazardous property?
A. Health
B. Flammability
C. Reactivity
D. Other

_____ 9. What does the yellow section of the *Emergency Response Guidebook* (ERG) list?
A. Chemicals by four-digit United Nations (UN) or North America (NA) number
B. Chemicals alphabetically
C. Emergency action guides
D. Initial isolation distances

_____ 10. A DOT placard with the number 6 in the bottom corner indicates what chemical hazard is being transported?
A. Explosives
B. Flammables
C. Poisons
D. Water reactives

_____ 11. Flammable gases that spontaneously ignite in air are what type gases?
A. Cryogenic gases
B. Oxidizing gases
C. Pyrophoric gases
D. Compressed gases

_____ 12. This class of flammable liquid includes flammable liquids with flashpoints below 73°F (22.8°C) and with boiling points at or above 100°F (37.8°C).
A. Class IA liquids
B. Class IB liquids
C. Class IC liquids
D. Class ID liquids

_____ 13. Of the various methods used to transport hazardous materials, which method is rarely and least likely to be involved in emergencies?
A. Rail
B. Highway
C. Waterway
D. Pipeline

_____ 14. This tanker is one of the most common chemical tankers and typically carries gasoline or other flammable and combustible materials.
A. MC 306
B. MC 307
C. MC 312
D. MC 331

_____ 15. These type tanks are both shipping and storage vessels.
A. Bulk tanks
B. Dewars
C. Intermodal
D. Nonbulk storage tanks

Fill in the Blank

Read each item carefully, and then complete the statement by filling in the missing word(s).

1. The classification system for flammable and combustible liquids is found in _____ _____, _____ _____ _____ _____ and is based on the division of flammable liquids into three main categories: class I liquids, class II liquids, and class III liquids.
2. Class II and class III liquids include all _____ liquids.

3. Toxic gases are labeled "__________ __________" for shipment.
4. A __________ __________ is defined in NFPA 55 as a gas not in solution that packaged under the charged pressure is entirely gaseous at a temperature of 68°F (20°C).
5. A __________ __________ is one that, at 68°F (20°C) inside its closed container, exists partly in the liquid state and partly in the gaseous state, and it is under pressure as long as any liquid remains in the container.
6. A __________ __________, as defined by the U.S. Department of Transportation (DOT), is a material that poses an unreasonable risk to the health and safety of operating emergency personnel, the public, and/or the environment if it is not properly managed and controlled during handling, storage, manufacture, processing, packaging, use and disposal, or transportation.
7. Labels are smaller versions (4" diamond-shaped indicators) of placards and are used on the four sides of individual boxes and smaller packages __________ __________.
8. __________ __________ diamonds are found on the outside of buildings, on doorways to chemical storage areas, and on fixed storage tanks.
9. Container storage of __________ __________ __________ __________ can be found in mercantile and industrial occupancies and in general-purpose and flammable liquid warehouses.
10. Specially designed storage cabinets are available for storing not more than __________ __________ __________ of class I, class II, and class IIIA liquids.

True/False

If you believe the statement to be more true than false, write the letter "T" in the space provided. If you believe the statement to be more false than true, write the letter "F."

_____ 1. The preferred method of dispensing flammable and combustible liquids from a drum is by use of an approved hand-operated pump drawing through the top.

_____ 2. In North America, there are three types of gas containers: cylinders, drums, and tanks.

_____ 3. The BLEVE hazard is restricted to containers of combustible gases, and the major cause of such BLEVEs in storage is improper usage.

_____ 4. An intermodal is any vessel or receptacle that holds material, including storage vessels, pipelines, and packaging.

_____ 5. It is good practice to eliminate sources of ignition in places where low flashpoint flammable liquids are stored, handled, or used, even though no vapor may ordinarily be present.

_____ 6. LP gas has a pungent chemical derived from the oil found in almonds added so that the presence of leaking gas can be detected by occupants before a dangerous level is reached.

_____ 7. A carboy is a glass, plastic, or steel container that holds 5 to 15 gallons of product.

_____ 8. Whenever possible in manufacturing processes involving flammable or combustible liquids, equipment such as compressors, stills, towers, and pumps should be locked in a closed room.

_____ 9. Pyrophorics pose a substantial threat if the Dewar fails to maintain the low temperature.

_____ 10. Fire protection safeguards for gas storage reflect the hazards of the container/gas combination and the hazards of the gas when it escapes from the container.

Vocabulary

Define the following terms using the space provided.

1. Bills of lading:

2. Cryogenic gas:

3. Dewar containers:

4. Material Safety Data Sheet (MSDS):

5. Pyrophoric gases:

Short Answer

Complete this section with short written answers using the space provided.

1. Identify the two classifications of gases according to NFPA 55, *Compressed Gases and Cryogenic Fluids Code*.

2. Identify the nine DOT chemical families recognized in the ERG.

3. Identify and briefly describe the four colored sections of the ERG:

4. Identify the topics an SDS generally includes.

5. Pesticide bags must be labeled with specific information. What information must be included on these bags?

Labeling

Label the following chemical tankers.

1. ______________________

2. ______________________

3. ______________________

4. ______________________

5. ______________________________

6. ______________________________

7. ______________________________

8. ______________________________

Fire Alarm

The following case scenario will give you an opportunity to explore the concerns associated with being a fire inspector. Read the scenario, and then answer each question in detail.

A chemical pipeline is running from a local chemical plant through your county, and many local fire departments are requesting information about it. You are the fire inspector responsible for pipeline inspection.

1. What information can you give the local fire departments about pipeline contents, ownership, and maintenance?

2. What information should be available to you and the local fire departments concerning emergencies with this pipeline?

This same chemical plant is in your jurisdiction and uses a variety of hazardous materials, many of them flammable.

3. How would you handle an inspection of the fire protection systems this chemical plant has in place and in service?

Safe Housekeeping Practices

The Workbook Activities

The following activities have been designed to help you. Your instructor may require you to complete some or all of these activities as a regular part of your fire inspector training program. You are encouraged to complete any activity that your instructor does not assign as a way to enhance your learning in the classroom.

Chapter Review

The following exercises provide an opportunity to refresh your knowledge of this chapter.

Matching

Match each of the terms in the left column to the appropriate definition in the right column.

_____	**1.** Defensible space	**A.** Upward or downward incline or slant, usually calculated as a percentage.
_____	**2.** Wildland/urban interface	**B.** Loose packing material (usually wood) protecting a ship's cargo from damage or movement during transport.
_____	**3.** Aspect	**C.** Any finely divided solid material that is 16.5 mil (420 µm) or smaller in diameter (material passing a U.S. No. 40 Standard Sieve) and presents a fire or explosion hazard when dispersed and ignited in air.
_____	**4.** Spontaneous ignition	**D.** A continuous progression of fuels that allows fire to move from brush to limbs to tree crowns or structures.
_____	**5.** High-piled storage	**E.** Devices installed above a cooking appliance to direct and capture grease-laden vapors and exhaust gases.
_____	**6.** Fuel ladder	**F.** Initiation of combustion of a material by an internal chemical or biological reaction that has produced sufficient heat to ignite the material.
_____	**7.** Slope	**G.** Solid-piled, palletized, rack storage, bin box, and shelf storage in excess of 12' (3.7 m) in height.
_____	**8.** Hood and exhaust system	**H.** Compass direction toward which a slope faces.
_____	**9.** Dunnage	**I.** An area as defined by the authority having jurisdiction (AHJ) (typically a width of 30' [9.14 m] or more) between an improved property and a potential wildland fire where combustible materials and vegetation have been removed or modified to reduce the potential for fire on improved property spreading to wildland fuels or to provide a safe working area for fire fighters protecting life and improved property from wildland fire.
_____	**10.** Combustible dust	**J.** Any area where wildland fuels threaten to ignite combustible homes and structures.

CHAPTER

13

Multiple Choice

Read each item carefully, and then select the best response.

_____ 1. Which of the following is *not* a basic requirement of good housekeeping?
- A. Equipment arrangement and layout
- B. Proper housekeeping tools and supplies
- C. Material storage and handling
- D. Operational neatness, cleanliness, and orderliness

_____ 2. Which of the following is not a poor housekeeping threat for the exterior of a structure?
- A. Obstructions to the site
- B. Obstructions to fire protection equipment
- C. Closed and locked doorways
- D. Fire exposure threats

_____ 3. The term used to describe posting of an address that can aid emergency responders in finding a building or residence is:
- A. premise identification.
- B. NFPA 1.
- C. property ID.
- D. house addressing.

_____ 4. NFPA 1 requires that dumpsters and similar waste receptacles should be located at least how many feet from combustible buildings?
- A. 10 feet
- B. 30 feet
- C. 100 feet
- D. There is no NFPA requirement

_____ 5. The minimum separation distance for 50 or more pallets is typically how many feet from buildings?
- A. 10 feet
- B. 30 feet
- C. 100 feet
- D. There is no NFPA requirement

_____ 6. What is the essential intent of a defensible space?
- A. Maintaining a clear distance around fire department connections (FDCs)
- B. Keeping entry gates opened
- C. Clearing all obstructions to the site
- D. Breaking the fuel ladder

_____ 7. Why would overgrown vegetation such as weeds, tall grass, brush, shrubs and trees concern a fire inspector?
- A. Wildland/urban interface issues
- B. Obstruct views
- C. Impair access to fire hydrants or fire department connections
- D. All of the above

_____ 8. Oily waste, oily rags, and oily towels can be the cause of what?
- A. Dust explosion
- B. Corrosive ignition
- C. Spontaneous ignition
- D. Endothermic reaction

_____ 9. According to NFPA 13, *Standard for the Installation of Sprinkler Systems*, aisles between racks of storage or solid piles on pallets should be a minimum of how wide?
- A. 2 feet
- B. 4 feet
- C. 10 feet
- D. 12 feet

_____ 10. NFPA 13 requires an aisle a minimum of how many inches wide between exterior walls of the building and materials that will absorb water and expand?
- A. 12"
- B. 24"
- C. 36"
- D. 48"

Fill in the Blank

Read each item carefully, and then complete the statement by filling in the missing word(s).

1. One of the more common problems seen by fire inspectors is storage of combustibles and carts in the _____ _____.
2. Minimum aisle dimensions and storage pile dimensions are based on a number of factors, including the commodity being stored and level of _____ _____ provided.
3. Emptying trash and waste at frequent intervals to prevent accumulation is one example of an _____ _____ _____.
4. Loose packing material, known as _____, is used to pack, support, and brace products within shipping containers, inside railcars, on flatbed trucks, and inside a ship's holds.
5. Businesses that handle flammable or combustible liquids should have an _____ _____ _____ for handling spills and leaks.
6. As part of a routine fire inspection, you should review or discuss the flammable liquid _____ _____ or procedures that the property uses.
7. In sprinkler-protected buildings, NFPA 1 and NFPA 13 require that storage be kept at least _____ below sprinklers.
8. A _____ _____ is a power-ventilated enclosure around a spraying operation or process that limits the escape of the material being sprayed and directs these materials to an exhaust system.
9. Grease accumulation on cooking hoods, on the hoods' grease filters, or inside the exhaust duct represents a serious _____ _____ _____.
10. _____ should be prohibited in areas where dust accumulations are present, such as woodworking plants, and in areas where there are combustible decorations.

True/False

If you believe the statement to be more true than false, write the letter "T" in the space provided. If you believe the statement to be more false than true, write the letter "F."

_____ 1. Correcting identified housekeeping issues is often a minor expense for the business or property owner.

_____ 2. Improperly stored materials can obstruct access to fire protection equipment, such as fire extinguishers, control valves, and fire alarm pull-stations.

_____ 3. NFPA 1 requires that combustible rubbish not stored in fire-rated rooms or vaults must be removed from the building by the property owner daily.

_____ 4. Fire inspectors need not be aware of hidden ash cans or cigarette butts in areas where smoking is prohibited.

_____ 5. Because of the low number of fires associated with spray finishing operations, the area or spray booth is not required to have an automatic fire suppression system.

_____ 6. A minimum of 18" (45.7 cm) clearance is specified in NFPA 54, *National Gas Fuel Code*, and NFPA 211, *Standard for Chimneys, Fireplaces, Vents, and Solid Fuel-Burning Appliances*, for gas and electric heaters and 36" (91.4 cm) clearance for high-heat producing appliances, such as boilers, incinerators, and solid-fuel burning appliances.

_____ 7. There are no restrictions on the amounts of certain types of compressed gases, such as flammable gases, permitted in an area.

_____ 8. Many storage and retail occupancies, sometimes referred to as bigbox stores, utilize high-piled storage arrangements to maximize space.

_____ 9. The use of flammable cleaning solvents is becoming fairly rare because of development of nonflammable solvents that have no flash points, are very stable, and have limited toxicity problems.

_____ 10. If a floor cleaning or refinishing operation is being conducted, ventilation is not an issue if materials having a flash point above the highest room temperature are used.

Vocabulary

Define the following terms with the space provided.

1. Aspect:

____

____

____

2. Dunnage:

____

____

____

3. Fuel ladder:

____

____

____

Short Answer

Complete this section with short written answers using the space provided.

1. Identify the four major fire and life safety objectives that safe housekeeping practices accomplish.

2. What can be the result of poor housekeeping practices outside of a building?

3. Identify and describe three basic requirements of good housekeeping.

4. Define a fuel ladder and explain how to break it.

5. Identify where dust and lint fire issues can exist, and what a fire inspector should be aware of while inspecting these industries.

Fire Alarm

The following case scenario will give you an opportunity to explore the concerns associated with being a fire inspector. Read the scenario, and then answer each question in detail.

You are doing a fire inspection of a recently opened bigbox store and notice most of the stock is accessible, but in "high-pile" storage.

1. How can you determine appropriate aisle widths and minimum allowed distances from exterior walls for some of the products such as toilet paper and paper towels?

2. How will you determine the minimum distance that material can be safely stored under the store's sprinkler system?

Writing Reports and Keeping Records

The Workbook Activities

The following activities have been designed to help you. Your instructor may require you to complete some or all of these activities as a regular part of your fire inspector training program. You are encouraged to complete any activity that your instructor does not assign as a way to enhance your learning in the classroom.

Chapter Review

The following exercises provide an opportunity to refresh your knowledge of this chapter.

Matching

Match each of the terms in the left column to the appropriate definition in the right column.

_____	**1.** Freedom of Information Act (FOIA)	**A.** Official notice of violation and statement of required corrective action.
_____	**2.** Building inspection reports	**B.** Written correspondence used when violations are not corrected to notify the owner that legal action may be taken to ensure code compliance.
_____	**3.** Fire code violation notification	**C.** A record of temporary permits authorizing the recipient to burn on a specific site for a specific period.
_____	**4.** Burning permits	**D.** Signed into law on July 4, 1966, by President Lyndon B. Johnson, it allows public access to government records.
_____	**5.** Final notice	**E.** May include records pertaining to the decommissioning of underground fuel tanks.
_____	**6.** Shall	**F.** Signed into law on October 17, 1986, by President Ronald Reagan. Provides the public with information about potential chemical hazards in their community.
_____	**7.** May	**G.** Indicates that you have the power to do something, but are not obligated to do so.
_____	**8.** Emergency Planning and Community Right-to-Know Act of 1986	**H.** Is mandatory and used when an action is required; no latitude or discretion is allowed.
_____	**9.** Is authorized to	**I.** Allows discretion; you are not necessarily required to perform the action.
_____	**10.** Property information	**J.** The portion of the Fire Inspection Report that describes the building itself.

Multiple Choice

Read each item carefully, and then select the best response.

_____ **1.** How long should fire inspection records be kept?

- **A.** 3 years
- **B.** 6 years
- **C.** 12 years
- **D.** As long as the building exists

CHAPTER

14

______ **2.** What font size is preferred when writing letters to a business or inspection site?

A. 8 point
B. 9 point
C. 12 point
D. 14 point

______ **3.** Which of the following writing errors could reflect poorly on you and your agency?

A. Improper grammar
B. Word usage
C. Poor spelling
D. All of the above

______ **4.** Which of the following will provide visual documentation of conditions observed during a fire inspection?

A. Sketches
B. Diagrams
C. Photographs
D. All of the above

______ **5.** What is the maximum length you should limit the letters written to companies you have inspected?

A. 3 paragraphs
B. 7 paragraphs
C. 2 pages
D. 4 pages

Fill in the Blank

Read each item carefully, and then complete the statement by filling in the missing word(s).

1. A ________________ occurs when a member of the public indicates that there is, in that person's opinion, a safety issue at a building.

2. If a violation is not dealt with correctly, ________________ must occur until the problem is corrected.

3. ________________ ________________ are the original on-scene description of conditions that existed at the time of the fire inspection.

4. A ________________ ________________ provides evidence of any fire and life safety hazards that you identified and the building's degree of code compliance.

5. Whenever possible, ________________ should be drawn to scale, or include dimensioning measurements and information, and use standard mapping symbols such as those found in NFPA 170, *Standard for Fire Safety and Emergency Symbols*.

6. ________________ are commonly used for final notices and reminders, simple inspection reports, or cover letters for more detailed inspection reports.

7. A ________________ ________________ should usually be reserved for times that the owner/occupant makes little or no effort to correct problems and deficiencies.

8. Most often, ________________ are used to respond to a request for information or code interpretation.

9. ____________________ are effective reminders of what you actually saw during your inspection, and they can often convey unsafe conditions better than words can.

10. Whenever you conduct a fire inspection, all documentation and correspondence is subject to use in a ____________________ ____________________.

True/False

If you believe the statement to be more true than false, write the letter "T" in the space provided. If you believe the statement to be more false than true, write the letter "F."

_____ 1. If a complainant is anonymous, a fire safety complaint should be ignored.

_____ 2. In many cases, a well-designed checklist is the only fire inspection record needed.

_____ 3. When drawing a sketch of a building during a fire inspection, everything must be to scale.

_____ 4. Verbal communications are superior to written records for documentation purposes.

_____ 5. All written communications are official and public documents, and state law requires you to keep them on file for a certain time before they can be discarded.

_____ 6. Reinspection is not a completely new inspection; if you identify additional problems you should not include them in the reinspection report as items separate from the original issues.

_____ 7. Inspection files are an important tool to improve fire safety and, if needed, can be used in legal proceedings.

_____ 8. The proper way to document a fire safety violation is to indicate both the violation and what must be done to correct it.

_____ 9. One way to overcome communication barriers is to practice writing the way you speak.

_____ 10. A fire inspection report can take several forms, from checklists to detailed reports.

Short Answer

Complete this section with short written answers using the space provided.

1. Identify six basic design elements each letter written should include:

2. What contact information should be included in reports for every property owner and tenant?

3. List four items that should be included when recording property information.

4. All fire inspection reports should contain what common elements?

5. Identify five common uses of written fire records.

Fire Alarm

The following case scenario will give you an opportunity to explore the concerns associated with being a fire inspector. Read the scenario, and then answer the question in detail.

Use your fire department fire inspection form and inspect your fire station. Be thorough and professional. Give your instructor a copy of the finished report to be reviewed for quality of work.

Life Safety and the Fire Inspector

The Workbook Activities

The following activities have been designed to help you. Your instructor may require you to complete some or all of these activities as a regular part of your fire inspector training program. You are encouraged to complete any activity that your instructor does not assign as a way to enhance your learning in the classroom.

Chapter Review

The following exercises provide an opportunity to refresh your knowledge of this chapter.

Matching

Match each of the terms in the left column to the appropriate definition in the right column.

_____	**1.** Indirect costs	**A.**	Personnel who teach fire safety messages to target audiences.
_____	**2.** Public information	**B.**	A national database that collects data about fire incidents including the collection of many of the underlying factors that caused the fire.
_____	**3.** Public education	**C.**	Expenses that must be paid but would otherwise not have occurred without the program such as printing, supplies, or fuel.
_____	**4.** Severity	**D.**	Activities that are focused on building a positive image of any organization.
_____	**5.** Fire and life safety educators	**E.**	Costs that the fire department would have incurred whether the fire and life safety education program existed or not, such as labor and apparatus costs.
_____	**6.** Public relations	**F.**	An individual or group that is impacted by an issue.
_____	**7.** National Fire Incident Reporting System (NFIRS)	**G.**	Evidence that may or may not be true, used to generalize when there is insufficient data to base it on.
_____	**8.** Stakeholder	**H.**	The amount of death, injury, or damage that is the result of an incident.
_____	**9.** Direct costs	**I.**	Information that is disseminated to the public via the fire department.
_____	**10.** Anecdotal evidence	**J.**	Teaching a safety message with the goal of reinforcing good behaviors, or changing undesirable behaviors, to make the community safer.

Multiple Choice

Read each item carefully, and then select the best response.

_____ **1.** A professional who is committed to saving lives, reducing injuries, and saving property is a:

A. public educator.
B. public relations.
C. human resources.
D. fire and life safety educator.

CHAPTER 15

______ **2.** Information about incidents that have occurred, events that will be taking place, or other matters of public concern is called:
A. public education.
B. public information.
C. public relations.
D. public exposure.

______ **3.** The term used to describe evidence that may or may not be true and is used to generalize when there is insufficient data is:
A. anecdotal evidence.
B. antidotal evidence.
C. antiquated evidence.
D. inadvertent evidence.

______ **4.** What should be your first step toward creating a fire and life safety education program?
A. Identify stakeholders
B. Develop goals and objectives
C. Develop relationships
D. Determine the level of commitment from the fire department

______ **5.** The presentation of a safety message with the sole goal of reinforcing good behaviors or changing undesirable behaviors to make the community safer is called:
A. public education.
B. public information.
C. public relations.
D. public exposure.

______ **6.** Which of the following can be an excellent source for fire safety program content and assistance in content development?
A. U.S. Fire Administration (USFA)
B. Commercial products
C. Other fire departments
D. All of the above

______ **7.** Activities that are focused on promoting the organization by creating a public perception that is positive and that creates public support for the organization are:
A. public education.
B. public information.
C. public relations.
D. public exposure.

______ **8.** Which organization is an excellent resource for data on fire prevention and life safety?
A. U.S. Federal Emergency Management Agency Learning Resource Center (FEMA LRC)
B. USFA
C. National Fire Protection Association (NFPA)
D. All are excellent resources

Fill in the Blank

Read each item carefully, and then complete the statement by filling in the missing word(s).

1. Proactive ________________ ________________ generally results in a positive public image of the fire department and also provides a safety message.
2. Virtually every fire department provides some form of public education; however, the effectiveness of these programs is difficult to prove without ________________ ________________.
3. ________________ ________________ include expenses that must be paid but would otherwise not have occurred without the program such as printing, supplies, and fuel.
4. Once data is collected, you will quickly see that some events occur more often than others, termed ________________, and that some of the problems are more significant when they do occur than others, termed ________________.
5. You can use ________________ ________________ to develop support and resources, and to help solve an identified problem.
6. ________________ ________________ are those costs that the fire department would have incurred whether the program existed or not.
7. Fire and life safety educators are personnel who teach ________________ ________________ messages to target audiences.
8. Often ________________ ________________ is used to determine the hazards in the community.
9. The state fire marshal amasses the data from the local fire departments as part of the ________________ ________________ ________________ ________________ ________________ ________________.
10. There are numerous sources of data available. The best choice of data is ________________ data.

True/False

If you believe the statement to be more true than false, write the letter "T" in the space provided. If you believe the statement to be more false than true, write the letter "F."

_____ 1. Once the most important stakeholders are identified, you must work to educate them about the problem and then encourage them to assist you in taking action.

_____ 2. A dilemma can occur with problems that are low frequency but high severity and those that are high frequency but low severity.

_____ 3. An example of indirect costs would be the value of fire personnel time spent presenting a program.

_____ 4. Objectives are broad and reflect the overall direction of the program.

_____ 5. Truly useful data provides just the basics: time, date, location, type of event, and who responded.

_____ 6. Providing high-quality public relations always provides effective public education.

_____ 7. FEMA's LRC has many research papers about the success of various fire and life safety education programs in fire departments across the country.

_____ 8. To meet the mission of saving lives and property, you may present programs on the importance of car seats, carbon monoxide detectors, or bicycle helmets.

_____ 9. One of the most important data sets available to a fire educator is the U.S. Census Data.

_____ 10. Virtually no fire department provides some form of public education because the effectiveness of these programs is difficult to prove.

Vocabulary

Define the following terms using the space provided.

1. Direct costs:

2. Fire and life safety educators:

3. National Fire Incident Reporting System (NFIRS):

4. Public relations:

Short Answer

Complete this section with short written answers using the space provided.

1. Identify 10 questions you should consider that will aid in the success of your fire and life safety education program.

2. How can you measure the effectiveness of a fire and life safety program?

3. Identify five stakeholders potentially available to you when addressing a local problem of children playing with matches and lighters.

4. What two main objectives should your fire and life safety programs address concerning the risks of fire?

Fire Alarm

The following case scenario will give you an opportunity to explore the concerns associated with being a fire inspector. Read the scenario, and then answer the question in detail.

You have researched and collected quite a bit of data identifying fire and safety issues in your jurisdiction. It is now time to use this data to assist you in developing your life safety programs.

1. What will assist you in determining a priority for your issues based on frequency and severity of the problems?

Answer Key

Chapter 1: Introduction to Fire Inspector

Matching

1. C (page 5)
2. E (page 4)
3. F (page 5)
4. H (page 7)
5. B (page 6)
6. A (page 5)
7. I (page 7)
8. D (page 5)
9. J (page 8)
10. G (page 6)

Multiple Choice

1. D (page 4)
2. A (page 6)
3. B (page 6)
4. C (page 7)
5. C (page 7)
6. B (page 7)
7. D (page 8)
8. A (page 8)
9. C (page 9)
10. C (page 8)

Fill in the Blank

1. fire inspection (page 4)
2. Fire investigators (page 5)
3. Fire Inspector II (page 6)
4. code enforcement system (page 7)
5. select for adoption (page 7)
6. job description (page 8)
7. notice of violation (page 8)
8. permit (page 8)
9. Ethical choices (page 9)

True/False

1. T (page 4)
2. F (page 4)
3. T (page 4)
4. T (page 5)
5. T (page 5)
6. F (page 5)
7. T (page 5)
8. T (page 6)
9. T (page 7)
10. F (page 8)

Vocabulary

1. **Code:** A standard that is an extensive compilation of provisions covering broad subject matter or that is suitable for adoption into law independently of other codes and standards. (page 6)
2. **Standard:** A document whose main text contains only mandatory provisions using the word "shall" to indicate requirements and is in a form generally suitable for mandatory reference by another standard or code or for adoption into law. Nonmandatory provisions shall be located in an appendix or annex, footnote, or fine print note and are not to be considered a part of the requirements of a standard. (page 7)
3. **Fire marshal:** A member of the fire department who inspects businesses and enforces laws that deal with public safety and fire codes. (page 5)
4. **Fire Inspector I:** An individual at the first level of progression who has met the job performance requirements specified in this standard for Level I. The Fire Inspector I conducts basic fire inspections and applies codes and standards (NFPA 1031). (page 5)

Short Answer

1. The job performance requirements a Fire Inspector I must meet, per NFPA 1031 are:
   - Inspect structures in the field and write reports based on observations and findings.
   - Identify the need for a permit and how to obtain the permit.
   - Recognize the need for a plan review and when to send it out to an expert for further review.
   - Investigate common complaints and act accordingly to resolve any compliance or safety issues.
   - When presented with a fire protection, fire prevention, or life safety issue, identify the applicable code or standard that is being violated.

- Participate in legal proceedings and provide testimony or written comments as required.
- Identify the occupancy classification of a single-use occupancy.
- Compute the allowable occupant load of a single-use occupancy, post that number publicly, and take corrective action if overcrowding occurs.
- Evaluate the exits of an existing building to ensure that a safe building evacuation may take place during an emergency.
- Verify the type of construction for an addition or remodeling project and direct compliance as needed.
- Determine if existing fixed fire suppression systems are working properly through testing and observation.
- Determine if existing fire detection and alarm systems are working properly through testing and observation.
- Determine if existing fixed portable fire extinguishers are working properly through testing and observation.
- Recognize any hazardous conditions involving equipment, processes, and operations in an occupancy.
- Compare an approved plan to an existing fire protection system and identifying and acting on any deficiencies found.
- Verify that emergency planning and preparedness measures are in place and have been practiced.
- Inspect the emergency access for the fire department to an existing site.
- Verify that hazardous materials, flammable and combustible liquids and gases, are stored, handled, and used in accordance with local codes and laws.
- Recognize a hazardous fire growth potential in a building or space.
- Determine code compliance, given the codes, standards, and policies of the jurisdiction and a fire protection issue.
- Verify that a building has a sufficient water supply in case of a fire. (pages 5, 6)

2. The job performance requirements a Fire Inspector II must meet, per NFPA 1031 are:

- Process a permit application.
- Process a plan review application.
- Investigate complex complaints, given a reported situation or condition, and bring the issue to a resolution.
- Recommend modifications to codes and standards of the jurisdiction based on a fire safety issue.
- Recommend policies and procedures for the delivery of inspection services, given management objectives.
- Compute the maximum allowable occupant load of multiuse building based on in-the-field observations or a description of its uses.
- Identify the occupancy classifications of a mixed-use building based on in-the-field observations or a description of its uses.
- Determine the building's area, height, occupancy classification, and construction type by examining an approved plan, a description of a building, or the construction features.
- Evaluate the fire protection systems and equipment to ensure they can protect the life safety of occupants from any hazards present in the structure.
- Analyze the elements of the exits in a building or portion of a building to ensure they meet with applicable codes and standards.
- Evaluate hazardous conditions involving equipment, processes, and operations in a building.
- Evaluate emergency planning and preparedness procedures using copies of existing or proposed plans and procedures.
- Verify code compliance for storage, handling, and use of hazardous materials, flammable and combustible liquids and gases.
- Determine the fire growth potential in a building or space based on the contents, interior finish, and construction elements.
- Inspect emergency access for the fire department to a site.
- Verify compliance with construction documents and ensure that life safety systems and building services equipment are installed, inspected, and tested to perform as described in the engineering documents and the operation and maintenance manuals.

- Classify the occupancy type of a building based on a set of plans, specifications, and a description of a building.
- Compute the maximum allowable occupant load in accordance with applicable codes and standards.
- Review the proposed installation of fire protection systems based on shop drawings and system specifications and ensure that the system is code compliant and installed in accordance with approved drawings.
- Verify that the means of exit elements are provided and meet all applicable codes and standards.
- Verify the construction type of a building based on a set of approved plans and specifications. (page 6)

3. Four of the more common fire positions a fire inspector may also assume are:
   - Fire marshal
   - Fire investigator
   - Fire and life safety education specialist
   - Fire protection engineer (page 5)
4. The path typically followed for codes and standards adoptions is:
   - The need to adopt a new code arises when a new code is created by a module code agency, a fire inspector expresses the need for a code to enforce, or a code needs modification.
   - The proposed modified language is prepared by technically competent staff or by consultants or the new code is collected.
   - The proposed modified code or new code is provided to elected officials for review.
   - Upon review and any corrections, the modified or new codes are posted for review by the general public.
   - A hearing is held to discuss the new or modified code by the elected officials with the general public.
   - Input is provided, considerations for changes are taken, modifications may be made, and the elected officials then vote the code into statute.
   - There may be an appeals process depending on the level of government enacting the regulations, state law governing adoption of safety regulations, and related oversight agencies such as state fire marshal's office, public utilities commission, and so on. (page 7)
5. Four types of permits are:
   - Use and occupancy permit: a permit to occupy a structure.
   - Special use permit: a permit to perform specified activities for a specified time period.
   - Fire protection equipment permit: a permit to install or use a particular fire protection equipment device.
   - Hazardous material use permit: a permit to use a specific product for a specific purpose for a specific time period. (page 8)

Fire Alarm

1. The key to improving ethical choices is to have clear organizational values. This can be accomplished by:
   - Having a code of ethics that is well known throughout the organization.
   - Selecting employees who share the values of the organization.
   - Ensuring that top management exhibit ethical behavior.
   - Having clear job goals.
   - Having performance appraisals that reward ethical behavior.
   - Implementing an ethics training program. (page 9)
2. The legal authority for fire inspection varies by state to state and, in many cases, varies within each state. The job description of the fire inspector should list the limits of legal authority that the position holds in the local jurisdiction and who/how to gain the next level of expertise or authority when needed.

   The range of authority of the fire inspector is defined by state or local law. (page 8)

Chapter 2: Building Construction

Matching

1. F (page 22)
2. E (page 25)
3. G (page 14)
4. H (page 27)
5. A (page 21)
6. I (page 26)
7. J (page 22)
8. D (page 14)
9. C (page 16)
10. B (page 26)

Multiple Choice

1. C (page 23)
2. B (page 23)
3. D (page 22)
4. A (page 20)
5. A (page 18)
6. C (page 19)
7. D (page 20)
8. D (page 23)
9. B (page 25)
10. A (page 27)
11. C (page 31)
12. A (page 16)

Fill in the Blank

1. Curtain walls (page 28)
2. Partition walls (page 28)
3. Parapet walls (page 28)
4. girder (page 29)
5. rafter (page 29)
6. criticality (page 29)
7. integrity (page 29)
8. annealed (page 32)
9. fire rated (page 33)
10. jalousie (page 33)

True/False

1. T (page 23)
2. F (page 34)
3. T (page 34)
4. T (page 34)
5. F (page 35)
6. T (page 35)
7. F (page 35)
8. T (page 36)
9. T (page 36)
10. F (page 30)

Vocabulary

1. **Pyrolysis:** The destructive distillation of organic compounds in an oxygen-free environment that converts the organic matter into gases, liquids, and char. (page 17)
2. **Parapet wall:** Walls on a flat roof that extend above the roofline. (page 28)
3. **Spalling:** Chipping or pitting of concrete or masonry surfaces. (page 15)
4. **Rafters:** Joists that are mounted in an inclined position to support a roof. (page 23)
5. **Dead load:** The weight of a building. It consists of the weight of all materials of construction incorporated into a building, including but not limited to walls, floors, roofs, ceilings, stairways, built-in partitions, finishes, cladding, and other similarly incorporated architectural and structural items, as well as fixed service equipment, including the weight of cranes. (page 22)

Short Answer

1. The most common building materials are wood, masonry, concrete, steel, aluminum, glass, gypsum board, and plastics. (page 14)
2. The key factors that affect the behavior of each of these materials under fire conditions are outlined here:

   Combustibility: Whether or not a material will burn determines its combustibility. Materials such as wood burn when they are ignited, releasing heat, light, and smoke, until they are completely consumed by the fire. Concrete, brick, and steel are noncombustible materials that cannot be ignited and are not consumed by a fire.

   Thermal conductivity: This characteristic describes how readily a material conducts heat. Heat flows very readily through metals such as steel and aluminum. By contrast, brick, concrete, and gypsum board are poor conductors of heat.

   Decrease in strength at elevated temperatures: Many materials lose strength at elevated temperatures. For example, steel loses strength and bends or buckles when exposed to fire temperatures, and aluminum melts in a fire. By contrast, bricks and concrete can generally withstand high temperatures for extended periods of time.

   Thermal expansion when heated: Some materials—steel, in particular—expand significantly when they are heated. A steel beam exposed to a fire stretches (elongates); if it is restrained so that it cannot elongate, it sags, warps, or twists.

As a general rule, a steel beam elongates at a rate of 1" (25.4 mm) per 10" (304 cm) of length at a temperature of 1000°F (538°C). (page 14)

3. Five different types of wood products commonly used in building applications today are:
   - *Solid lumber* is squared and cut into uniform lengths. Examples of solid lumber include the heavy timbers used in churches, mills, and barns and the lightweight boards used for siding and decorative trim.
   - *Laminated wood* consists of individual pieces of wood glued together. Lamination is used to produce beams that are longer and stronger than solid lumber and to manufacture curved beams.
   - *Wood panels* are produced by gluing together thin sheets of wood. Plywood is the most common type of wood panel used in building construction. Small chips (chipboard) or particles of wood (particleboard) can also be used to make wood panels, although these panels are usually much weaker than those constructed from plywood or solid lumber.
   - *Wood trusses* are assemblies of pieces of wood or wood and metal combinations; they are often used to support floors and roofs. The structure of a truss enables a limited amount of material to support a heavy load.
   - *Wooden beams* are efficient load-bearing members assembled from individual wood components. The shape of a wooden I-beam or box beam enables it to support the same load that a solid wood beam of its size could support. (page 17)
4. Five types of building construction are:
   - *Type I construction* (also referred to as fire-resistive construction) is the most fire-resistive category of building construction.
   - *Type II construction* is also referred to as noncombustible construction. All of the structural components in a Type II building must be made of noncombustible materials. The fire-resistive requirements, however, are less stringent for Type II construction than for Type I construction.
   - *Type III construction* is also referred to as ordinary construction because it is used in a wide variety of buildings, ranging from commercial strip malls to small apartment buildings.
   - *Type IV construction* is also known as heavy timber construction. A heavy timber building has exterior walls that consist of masonry construction and interior walls, columns, beams, floor assemblies, and roof structure that are made of wood.
   - In *Type V construction*, all of the major components are constructed of wood or other combustible materials. Type V construction is often called wood-frame construction and is the most common type of construction used today. (page 18)
5. The five NFPA 80 designations for fire doors and fire windows are:
   - Class A: Openings in fire walls and in walls that divide a single building into fire areas
   - Class B: Openings in enclosures of vertical communications through buildings and in 2-hour-rated partitions providing horizontal fire separations
   - Class C: Openings in walls or partitions between rooms and corridors having a fire-resistance rating of 1 hour or less
   - Class D: Openings in exterior walls subject to severe fire exposure from outside the building
   - Class E: Openings in exterior walls subject to moderate or light fire exposure from outside the building (page 35)

Fire Alarm

1. Heavy timber construction, if constructed to meet code, has no concealed spaces (voids). This structure helps reduce the horizontal and vertical fire spread that often occurs in ordinary construction buildings. Unfortunately, many heavy timber buildings do have vertical openings for elevators, stairs, or machinery, which can provide a path for a fire to travel from one floor to another.

   The solid wood columns, support beams, floor assemblies, and roof assemblies used in heavy timber construction withstand a fire much longer than the smaller wood members used in ordinary and lightweight combustible construction. (page 20)
2. Once involved in a fire, the structure of a heavy timber building can burn for many hours. A fire that ignites the combustible portions of heavy timber construction is likely to burn until it runs out of fuel and the building is reduced to a pile of rubble. As the fire consumes the heavy timber support members, the masonry walls become unstable and collapse. (page 20)

Chapter 3: Types of Occupancies

Matching

1. I (page 42)
2. D (page 44)
3. G (page 52)
4. H (page 51)
5. F (page 47)
6. J (page 49)
7. B (page 50)
8. A (page 45)
9. C (page 47)
10. E (page 43)

Multiple Choice

1. B (page 46)
2. C (page 48)
3. A (page 43)
4. B (page 48)
5. D (page 52)
6. A (page 49)
7. C (page 45)
8. D (page 46)
9. B (page 43)
10. C (page 52)
11. A (page 44)
12. B (page 48)

Fill in the Blank

1. intended (page 42)
2. occupancy classifications (page 42)
3. not required (page 42)
4. fire protection (page 43)
5. rooming houses (page 44)
6. different (page 45)
7. healthcare occupancy (page 46)
8. ambulatory (page 46)
9. daycare (page 47)
10. mercantile (page 49)

True/False

1. T (page 43)
2. F (page 44)
3. F (page 43)
4. T (page 45)
5. T (page 46)
6. T (page 46)
7. F (page 48)
8. F (page 48)
9. T (page 49)
10. F (page 47)

Vocabulary

1. **Occupancy:** The intended use of a building. (page 42)
2. **Mixed occupancy:** A multiple occupancy where the occupancies are intermingled. (NFPA 101, *Life Safety Code*). (page 52)
3. **Multiple occupancy:** A building or structure in which two or more classes of occupancy exist. (NFPA 101, *Life Safety Code*). (page 52)

Short Answer

1. According to the NFPA 101, *Life Safety Code,* there are 15 specific occupancy groupings:
   1. One- and two-family dwellings
   2. Lodging or rooming houses
   3. Hotels and dormitories
   4. Apartment buildings
   5. Residential board and care
   6. Health care
   7. Ambulatory health care
   8. Daycare
   9. Educational
   10. Business
   11. Industrial
   12. Mercantile
   13. Storage
   14. Assembly
   15. Detention and correctional (page 42)

2. Examples of residential board and care occupancies include:
   - Group housing arrangement for physically or mentally handicapped persons who normally attend school in the community, attend worship in the community, or otherwise use community facilities
   - Group housing arrangement for physically or mentally handicapped persons who are undergoing training in preparation for independent living, for paid employment, or for other normal community activities
   - Group housing arrangement for the elderly that provides personal care services, but that does not provide nursing care
   - Facilities for social rehabilitation, alcoholism, drug abuse, or mental health problems that contain a group housing arrangement and that provide personal care services but do not provide acute care
   - Assisted living facilities
   - Other group housing arrangements that provide personal care services but not nursing care (page 45)
3. Ambulatory healthcare occupancy services include one or more of the following:
   - Treatment for patients that renders the patients incapable of taking action for self-preservation under emergency conditions without the assistance of others.
   - Anesthesia that renders the patients incapable of taking action for self-preservation under emergency conditions without the assistance of others.
   - Emergency or urgent care for patients who, due to the nature of their injury or illness, are incapable of taking action for self-preservation under emergency conditions without the assistance of others. (page 46)
4. The three subclasses of daycare occupancies are:
   - Daycare occupancies
   - Group daycare homes
   - Family daycare homes (page 47)
5. There are five categories under detention and correctional occupancy that correspond to the degree of restraint of occupants within the facility. These categories are:
   - Use Condition I, Free Egress: Occupants are permitted to move freely to the exterior.
   - Use Condition II, Zoned Egress: Occupants are permitted to move from sleeping areas to other smoke compartments.
   - Use Condition III, Zoned-Impeded Egress: Occupants are permitted to move within any smoke compartment.
   - Use Condition IV, Impeded Egress: Occupants are locked in their rooms or cells, but they can be remotely unlocked.
   - Use Condition V, Contained: Occupants are locked in their rooms or cells with manually operated locks that can only be opened by facility staff members. (page 51)

Fire Alarm

1. With mixed occupancies, the means of egress, facilities, types of construction, protection, and other safeguards in the building shall comply with the most restrictive fire and life safety requirements of the occupancies involved. (page 52)
2. Guidance on the requirements for special structures can be found in NFPA 101 and other relevant standards. When inspecting a special structure, you should first determine the general occupancy classification and then determine the special requirements that apply. Occupancies in special structures, although often not easy to inspect, must be inspected in accordance with all pertinent requirements. (page 52)

Chapter 4: Fire Growth

Matching

1. D (page 68)
2. H (page 61)
3. J (page 69)
4. I (page 69)
5. A (page 60)
6. B (page 58)
7. E (page 58)
8. C (page 61)
9. G (page 69)
10. F (page 69)

Multiple Choice

1. C (page 63)
2. B (page 63)
3. A (page 63)
4. E (page 63)
5. D (page 63)
6. B (page 59)
7. A (page 59)
8. C (page 59)
9. D (page 74)
10. B (page 67)
11. B (page 58)
12. D (page 68)

Fill in the Blank

1. BLEVE (boiling liquid/expanding vapor explosion) (page 70)
2. decay phase (page 65)
3. fire triangle (page 60)
4. Flameover (page 68)
5. flame point (page 69)
6. fully developed (page 65)
7. gas (page 58)
8. growth phase (page 65)
9. Hypoxia (page 61)
10. Radiation (page 62)

True/False

1. T (page 59)
2. T (page 59)
3. F (page 58)
4. F (page 69)
5. F (page 61)
6. T (page 60)
7. T (page 58)
8. F (page 63)
9. F (page 65)
10. T (page 60)

Vocabulary

1. **Endothermic:** Reactions that absorb heat or require heat to be added. (page 59)
2. **Chemical energy:** Energy that is created or released by the combination or decomposition of chemical compounds. (page 59)
3. **Fire tetrahedron:** A geometric shape used to depict the four components required for a fire to occur: fuel, oxygen, heat, and chemical chain reactions. (page 60)
4. **Vapor density:** The weight of an airborne concentration (vapor or gas) as compared to an equal volume of dry air. (page 69)
5. **Volatility:** The ability of a substance to produce combustible vapors. (page 69)

Short Answer

1. Fires grow and spread by three primary mechanisms: conduction, convection, and radiation.
   - **Conduction** is the process of transferring heat through matter by movement of the kinetic energy from one particle to another.
   - **Convection** is the circulatory movement that occurs in a gas or fluid with areas of differing temperatures owing to the variation of the density and the action of gravity.
   - **Radiation** is the transfer of heat through the emission of energy in the form of invisible waves. (page 61)
2. Almost all of the gases produced by a fire are toxic to the body, including carbon monoxide, hydrogen cyanide, and phosgene. (page 60)
3. Although there are many variations on the methods used to extinguish fires, they boil down to four main methods: cooling the burning material, excluding oxygen from the fire, removing fuel from the fire, and interrupting the chemical reaction with a flame inhibitor. (page 62)
4. Fires are generally categorized into one of five classes: Class A, Class B, Class C, Class D, and Class K.
   - Class A fires involve solid combustible materials such as wood, paper, and cloth.
   - Class B fires involve flammable or combustible liquids such as gasoline, kerosene, diesel fuel, and motor oil.
   - Class C fires involve energized electrical equipment.
   - Class D fires involve combustible metals such as sodium, magnesium, and titanium.
   - Class K fires involve combustible cooking oils and fats in kitchens. (page 63)

5. Solid-fuel fires progress through four phases: the ignition phase, the growth phase, the fully developed phase, and the decay phase.
   - The ignition phase, in which the fire is limited to its point of origin, begins as a lighted match is placed next to a crumpled piece of paper. The heat from the match ignites the paper, which sends a small plume of fire upward. The heat generated from the paper sets up a small convection current, and the flame produces a small amount of radiated energy. The combination of convection and radiation serve to heat the fuel around the paper.
   - The growth phase, when the fire spreads to nearby fuel, occurs as the kindling starts to burn, increasing the convection of hot gases upward. The hot gases and the flame both act to raise the temperature of the wood located above the kindling. Energy generated by the growing fire starts to radiate in all directions. The convection of hot gases and the direct contact with the flame cause major growth to occur in an upward direction.
   - The fully developed phase produces the maximum rate of burning. All available fuel has ignited and heat is being produced at the maximum rate. At the fully developed phase, thermal radiation extends in all directions around the fire.
   - The final phase of a fire is the decay phase, the period when the fire is running out of fuel. During the decay phase, the rate of burning slows down because less fuel is available. The rate of thermal radiation decreases. (pages 64–65)

Fire Alarm

1. A wide variety of materials and hidden building elements contribute greatly to rapid fire growth. For example, batt (or paneled) insulation laid in ceilings must be kept free of light fixtures because the heat from the fixture can ignite the paper vapor seal, igniting a fire inside of the ceiling.
   - Combustible fiberboard is commonly used as insulating sheathing on wood frame buildings. It is also used as soundproofing. This material can support a fire hidden in the walls.
   - Foamed-plastic insulation is also used as sheathing, concealed in cavity walls, and glued to the interior surface of masonry wall panels.
   - Air-duct insulation, commonly installed years ago, was usually made of a hair felt with a high flame spread. (page 70)
2. There are three ways in which interior finishes increase a fire hazard:
   - They increase fire extension by surface flame spread.
   - They generate smoke and toxic gases.
   - They may add fuel to the fire, contributing to flashover. (page 71)

Chapter 5: Performing an Inspection

Matching

1. C (page 79)
2. E (page 80)
3. A (page 80)
4. D (page 79)
5. I (page 86)
6. A (page 80)
7. F (page 79)
8. G (page 83)
9. B (page 83)
10. H (page 83)

Multiple Choice

1. D (page 79)
2. A (page 79)
3. C (page 80)
4. A (page 81)
5. B (page 79)
6. C (page 83)
7. D (page 87)
8. D (page 89)
9. D (page 94)
10. B (page 89)
11. A (page 80)

Fill in the Blank

1. safe (page 79)
2. As-built diagrams (page 80)
3. legally (page 81)
4. site plan (page 87)
5. predetermined and structured (page 88)
6. combustibles (page 89)
7. contents hazard (page 90)
8. interior finishes (page 92)
9. flame rating (page 92)
10. preplan (page 92)

True/False

1. T (page 86)
2. T (page 88)
3. T (page 89)
4. F (page 92)
5. F (page 92)
6. T (page 91)
7. T (page 92)
8. T (page 93)
9. F (page 93)
10. T (page 95)

Vocabulary

1. **Business license or change of occupancy inspections:** Inspections that occur when the building department is notified of a new business requesting permission to open. (page 80)
2. **Complaint inspections:** Inspections that occur when someone registers a concern of a possible code violation. (page 79)
3. **Complaint form:** Form that lists in detail any complaint that is lodged with the fire inspection agency and is investigated. (page 83)

Short Answer

1. The inspection form should consist of the following elements:
   - A section showing the business name, address, and phone number.
   - A section showing the date of the inspection.
   - A section showing the area being inspected if the complex is large, if there are multiple buildings, or you are just inspecting one specific section.
   - A section for listing any code violations with the specific code citations.
   - A section for listing the reinspection date to remind the owner when you will be returning to check on code compliance.
   - Areas for the signatures of the fire inspector and a representative of the owner. It is a good idea to also have a place where the names could be printed beneath the signatures.
   - A section with a short legal statement stating the fire inspector's authority for the fire inspection is optional.
   - The legal statement should be reviewed by your agency's legal staff. (page 83)
2. Some of the more common housekeeping safety issues include:
   - Storing stock too close to the sprinkler heads
   - Garbage cans overflowing
   - An accumulation of materials in furnace and water heater rooms
   - Storing boxes and mops too close to pilot lights on furnaces and water heaters
   - Storing flammable materials incorrectly (page 86)
3. There are a few hazard violations that routinely appear, including:
   - **Electrical:** Electrical cords cannot be spliced; circuit breakers must be identified; extension cords cannot be used in place of permanent wiring, openings are not allowed in electrical panels; clear access to the electric panel must be maintained; cover plates are needed for junction boxes, switches, and outlets; and no multiplug adapters are allowed.
   - **Exit/emergency lights:** Lights must function properly and must not be obstructed.
   - **Exiting:** Exit doors must be operational and unobstructed, doors must close and latch but may not use deadbolt locks, storage is not allowed in halls or stairwells, exit doors must swing outward, and exit signage is required.
   - **Fire extinguishers:** Fire extinguishers must have current inspection tags and a minimum of 2A10BC rating; extinguishers must be mounted properly, unobstructed, and operational, indicated by proper signage where appropriate; extinguishers should be properly spaced for minimal travel distance; they should be of the proper type for the hazard; and there should be enough extinguishers to comply with fire codes.
   - **Fire detection and suppression systems:** This equipment must be accessible and kept in a normal status, fire department connections must be capped and accessible, storage cannot be too close to sprinklers, and rooms must be properly labeled. Specialized annual inspections are required on these systems.
   - **General:** A key box containing proper keys is required, good housekeeping must be maintained, address numbers on the building must be visible, high-pressure cylinders must be secured to the wall, gas meters must be protected, fire hydrants must be visible and accessible, no excessive amounts of flammable liquids may be stored, flammable

materials must be kept in the proper containers, ashtrays should be provided in smoking areas, no smoking signs must be provided where smoking is not permitted, emergency vehicle access must be unobstructed, and fire lanes must be identified.

- **Heating appliances:** All combustibles must be kept 36" (914 mm) away from a heat source. The heating appliance must be in good repair and easily accessible.
- **Openings:** All pipe chases must be filled, ceiling tiles must be intact and in place, any holes in drywall must be patched, any openings in fire walls must be repaired, and fire doors must work properly. (page 89)

4. Look at building features for the following items:
   - Check the means of egress including stairway enclosures.
   - Make certain that doors swing in the right direction.
   - Travel distances and the common paths of travel are not exceeded.
   - Doors close and latch.
   - Exit doors are not locked.
   - Fire doors should close and latch. If there are automatic devices used to operate and release the door, those pieces must function smoothly.
   - Smoke evacuation systems and systems designed to pressurize a stairwell or floor are complicated and should be tested by a professional company.
   - Ceiling tiles missing from the area around the sprinkler head are a deterrent to its operation.
   - Exit lighting requires looking to see if the lights are illuminated.
   - Emergency lighting requires some type of testing. In most cases this can be as easy as pushing the test button to see if the lights illuminate. It gets more difficult when those lights are more than 15' (457 cm) off the floor. Oftentimes those units have a separate circuit breaker and the lights come on simply by turning it off. Never do that yourself. Let the building owner do that because you will have no idea what else may be tied into the breaker. There have been instances where a breaker was turned off and computers also shut down, causing the loss of many hours of work.
   - More mundane features such as carpeting and interior finishes must also be checked for compliance. In certain public buildings, the furniture needs to have a tag indicating it will not promote flame spread. (page 90)
5. There are a number or basic or routine inspections that a fire inspector must perform. These include annual inspections, reinspections, complaint inspections, construction or final inspections, business license or change of occupancy inspections, and self-inspection. (page 79)

Fire Alarm

1. Knowledge can be gained from reading books and attending classes, but it cannot stop there. As you progress in the career of a fire inspector, new ideas, equipment, and trends emerge. It is your responsibility to stay abreast of the changes. The easiest way is to talk to contractors who should be experts in their field. If you show a willingness to learn, most are eager to help. You can also call manufacturers. Tell them that you would like some additional information about certain equipment or processes.

   One of the best ways to stay informed is to join a local fire inspectors association. Here common problems are discussed, vendors talk about new products, and there is the opportunity to meet other fire inspectors. The ability to call another fire inspector to discuss a problem is invaluable. (page 95)
2. Gaining experience takes time. The more you go into the same buildings, the more familiar you will become, and the quicker you can conduct the inspection. Additionally, the personnel at the building will be accustomed to seeing you and will know what to expect. Although still being professional, the initial formality begins to become a little more informal and relaxed. Combining knowledge and experience will make you a more confident fire inspector. You will also be able to see the building and possible violations on a different level than just black and white. (page 95)

Chapter 6: Reading Plans

Matching

1. F (page 103)
2. G (page 113)
3. I (page 106)
4. B (page 113)
5. J (page 107)
6. H (page 105)
7. E (page 106)
8. C (page 113)
9. D (page 104)
10. A (page 113)

Multiple Choice

1. A (page 107)
2. B (page 108)
3. B (page 107)
4. C (page 112)
5. C (page 107)
6. B (page 108)
7. D (page 107)
8. A (page 109)
9. C (page 109)
10. B (page 113)

Fill in the Blank

1. built (page 112)
2. plan review (page 103)
3. Structural plans (page 105)
4. site plan (page 109)
5. occupancy (page 109)
6. damper (page 109)
7. HVAC (page 110)
8. sprinklers (page 110)
9. detection (page 111)
10. deficiencies (page 111)

True/False

1. T (page 112)
2. T (page 103)
3. F (page 103)
4. T (page 104)
5. F (page 104)
6. F (page 105)
7. T (page 106)
8. F (page 106)
9. T (page 112)
10. T (page 113)

Vocabulary

1. **Architectural plan:** A drawing showing floor plans, elevation drawings, and features of a proposed building's layout and construction. (page 105)
2. **Mechanical plans:** Drawings in a plan set showing the proposed plumbing, HVAC, or other mechanical systems for a building. (page 106)
3. **Plan view:** A view on a drawing where a horizontal slice is made in the building or area and everything above or below the slice is shown. (page 107)
4. **Sectional view:** On a drawing in a plan set, a vertical slice of a building showing the internal view of the building. (page 107)
5. **Site plan:** A drawing showing the building and surrounding area, including items such as roads, driveways, and hydrants. (page 108)

Short Answer

1. The title block contains basic information about the drawing and includes some or all of the following information:
   - Date
   - Drawing number
   - Job number
   - Revision number or date
   - Project name and location
   - Designer's name or initials
   - Company information for the architectural or engineering firm
   - Architect or engineer's stamp and signature
   - Scale
   - Legend (page 104)

2. The code analysis should list the applicable building and fire codes for the project and based on those codes should show the following information:
   - **Occupancy classification:** The adopted model building code and NFPA 101, *Life Safety Code* may have different classifications for the occupancy, and both should be shown in the plan set. If there are multiple occupancies within the building, they should all be listed. If the occupancies will be separated, the hourly rating of the fire-resistance rated separation should be included.
   - **Construction type:** The construction type of the proposed building is based on NFPA 220, *Standard on Types of Building Construction*, or the adopted model building code classification. Chapter 2, *Building Construction*, discusses the five types of building construction in detail.
   - **Building area:** This shows the height and area for the proposed building and should also show the height and area limitations from the building or fire codes for the occupancy type. The building code, *Life Safety Code*, and local and state zoning laws may dictate the maximum area per floor and number of potentially occupied levels that a specific occupancy type can have.
   - **Occupant load:** Each area of the building will have an occupant load based on the minimum number of occupants that the egress system will need to support. The total for the building will be shown in this portion of the code analysis. The occupant load factors come from NFPA 101, Table 7.3.1.2. Occupant load factors and how to calculate the occupant load is discussed in detail in Chapter 7 *Occupancy Safety and Evacuation Plans*.
   - **Fire protection systems:** Most buildings built today have some form of a fire protection system ranging from an elaborate smoke evacuation system in a stadium to smoke alarms in a single-family residence. Sprinklers, suppression systems, and fire alarm systems may all be required by applicable building and fire codes. The code analysis should indicate whether these systems will be included and also whether or not they are required.
   - **Egress:** The code analysis often includes information about the egress arrangement. This may consist of travel distance, common path of travel, dead-end corridors, number of exits, remoteness, and accessibility of the means of egress. Again, the code requirements and building information should be provided. (page 104)
3. There are several types of drawings you will encounter in a plan set.
   - **Site plans:** A site plan is a drawing that shows the overview of the lot being reviewed.
   - **Structural plans:** The structural plan is a drawing showing the proposed building's load-bearing components.
   - **Architectural plans:** Architectural plans show the floor plans, reflected ceiling plans, building sections, elevations, and the building details.
   - **Electrical plans:** Electrical plans show circuits, outlets, and lighting within the proposed structure.
   - **Mechanical plans:** Mechanical plans show the heating, ventilation, and air-conditioning layouts of the proposed building. (page 105)
4. There are four basic types of views that will be encountered in a plan set.
   - **Plan view:** A view where a horizontal slice is made in the building or area and everything above or below the slice is shown.
   - **Elevation view:** Elevations show the exterior of the building and are labeled either using the direction the drawing is facing or by front, rear, left, and right.
   - **Sectional view:** A vertical slice of a building showing the internal view.
   - **Detail view:** Views of a specific element of construction or building feature in a larger scale providing more clarity. (page 107)
5. Most of the model codes and standards have provisions for equivalencies and alternatives.
   - **Equivalencies** set forth the permissibility of a method, process, system, or device of superior quality, strength, durability, and so on. It is important to note that the submitter is required to provide technical documentation to determine the equivalency and to show that the equivalency is approved for the intended purpose. If you are installing a sprinkler system with a pipe that is not of the types listed in NFPA 13, as long as there is technical documentation showing its durability and suitability for sprinkler system service it would be allowed under this clause, subject to approval of the authority having jurisdiction.
   - An **alternative clause** allows for the code provisions to be altered to an alternative that would not reduce the level of safety within the building. In many cases the use of alternatives in existing buildings may actually result in a safer building than was originally being occupied. (page 113)

Fire Alarm

1. If the approved plan set is not being used, you should halt construction until approval is obtained. This can be accomplished by following the procedures set up by your agency. Ordinances, laws, or other regulations will have set up the process for stopping work if the work has been misrepresented or is not being conducted in accordance with the approved plan set. This may seem excessive and will usually not be well received by the project manager, but it is a necessary step in ensuring that what is being built meets standards. You should explain that it is far easier to halt operations temporarily than it is to rebuild the project. (page 112)
2. When a proposed design, operation, process, or new technology exceeds your capabilities, the NFPA 1 allows for the use of technical assistance. This technical assistance involves the review of the matter in question by an independent third party. The owner or designer is usually responsible for the cost of the third-party review, so it is important that you quickly inform the submitter when a third-party review will be mandated. The evaluation performed by the third party review will be submitted to you for final approval. If you need some direction in the technical assistance process, consult *Guidelines for Peer Review in the Fire Protection Design Process* published by the Society of Fire Engineers. (page 113)

Chapter 7: Occupancy Safety and Evacuation Plans

Matching

1. C (page 146)
2. F (page 135)
3. D (page 136)
4. J (page 146)
5. B (page 146)
6. G (page 146)
7. H (page 124)
8. A (page 146)
9. I (page 124)
10. E (page 146)

Multiple Choice

1. A (page 119)
2. B (page 120)
3. A (page 124)
4. C (page 124)
5. C (page 125)
6. D (page 125)
7. A (page 125)
8. D (page 126)
9. B (page 129)
10. C (page 133)

Fill in the Blank

1. natural path of travel (page 133)
2. egress capacity (page 134)
3. Maximum egress capacity (page 134)
4. *Life Safety Code* (page 134)
5. occupant load (page 135)
6. horizontal exit (page 135)
7. Exit stairs (page 136)
8. Smoke-proof enclosures (page 136)
9. exit passageway (page 137)
10. poor substitute (page 137)

True/False

1. T (page 123)
2. F (page 124)
3. T (page 124)
4. F (page 127)
5. T (page 129)
6. T (page 136)
7. T (page 137)
8. F (page 138)
9. T (page 139)
10. T (page 140)

Vocabulary

1. **Area of refuge:** An area that is either (1) a story in a building where the building is protected throughout by an approved, supervised automatic-sprinkler system and has not less than two accessible rooms or spaces separated from each other by smoke-resisting partitions, or (2) a space located in a path of travel leading to a public way that is protected from the effects of fire, either by means of separation from other spaces in the same building or by virtue of location, thereby permitting a delay in egress travel from any level (NFPA 101). (page 146)
2. **Exit discharge:** That portion of a means of egress between the termination of an exit and a public way (NFPA 101). (page 124)
3. **Horizontal exit:** An exit between adjacent areas on the same deck that passes through an A-60 Class boundary that is contiguous from side shell to side shell or to other A-60 Class boundaries (NFPA 301). (page 146)

4. **Means of egress:** A continuous and unobstructed way of exit travel from any point in a building or structure to a public way, consisting of three separate and distinct parts: (1) the exit access, (2) the exit, and (3) the exit discharge. A means of egress comprises the vertical and horizontal travel and includes intervening room spaces, doorways, hallways, corridors, passageways, balconies, ramps, stairs, enclosures, lobbies, escalators, horizontal exits, courts, and yards (NFPA 101). (page 146)

Short Answer

1. The *Life Safety Code* forms the basis for most of the egress requirements contained in the model building and fire codes. The *Life Safety Code* also addresses the construction, protection, and occupancy features necessary to minimize dangers to life from the effects of fire, including smoke, heat, and toxic gases. It also establishes minimum criteria for the design of egress facilities to allow for the prompt escape of occupants from buildings or to safe areas within buildings. (page 119)
2. Occupant load is based on use of a space, not the occupancy. A school is an educational occupancy, but it has spaces of classrooms, kitchens, libraries, laboratories, swimming pools, and so on. Occupant load is determined by the nature of the use of a building or space and the amount of space available for that use. Because different generic uses are characterized by different occupant densities, the *Life Safety Code* has established occupant load factors for each use. (page 122)
3. A means of egress consists of three separate and distinct parts:
   - **Exit access:** That portion of a means of egress that leads to the entrance of an exit. In other words, the exit access is the travel anywhere within the building to the exit from the building. This may include travel within corridors, on stairs, or traversing open floor areas. There are limits an occupant can travel within the exit access to actually reach an exit, which is called the "maximum travel distance."
   - **Exit:** That portion of a means of egress that is between the exit access and the exit discharge. An exit may be composed of vertical and horizontal means of travel. For example, entering an exit stairway on the fourth floor would end the exit access for that floor and begin the exit even though there are many floors of stairs to walk until the exit discharge is reached.
   - **Exit discharge:** That portion of a means of egress between the termination of the exit and a public way. In other words, the exit discharge is the area between the exit and the nearest public way. All exits must terminate at a public way. Although the requirements for exit discharge are vaguely defined, the entire distance must be identifiable, reasonably direct, and essentially unimpeded. This could include such things as removing the accumulation of snow and ice during the winter, making certain the terrain is even, and removing obstacles that might hinder movement to the public way. Most designs exit onto a walkway that leads to the public way. (page 124)
4. The length of the exit access establishes the travel distance to an exit, an extremely important feature of a means of egress, because an occupant might be exposed to fire or smoke during the time it takes to reach an exit. (page 144)
5. To calculate the egress capabilities of a door, divide the door width by the factor 0.2.

   To calculate the width of the stairs, multiply the occupants by the factor 0.3. (page 127)

Fire Alarm

1. Required stair width is determined by the required egress capacity of each floor the stair serves, considered independently. It is not necessary to accumulate occupant loads from floor to floor to determine stair width. Each story or floor is considered separately when calculating the occupant load to be served by the means of egress from that floor. The size or width of the stair need only accommodate the portion of the floor's occupant load assigned to that stair. However, in a multistory building, the floor requiring the greatest egress capacity dictates the minimum stair width from that floor down to the exit discharge, in the direction of egress travel. It is not permissible to reduce such stair width along the remainder of the stairs encountered in traveling to the level of exit discharge, that is, stairs encountered in the direction of egress travel. Exits serving floors above the floor of greatest egress capacity are permitted to use egress components sized to handle the largest demand created by any floor served by that section of stair run. (page 127)
2. Standard occupant load factors are listed in Table 7.3.1.2 in the *Life Safety Code*. The table shows two types of occupant load factors: gross and net. Most of the factors shown in the table are gross factors. The gross factor is calculated for the building as a whole and measured from wall to wall. The net factor is only the space that can be occupied. It is the gross figure minus any tables, columns, or other unusable space. Net factors are used in occupancies where there will be higher concentrations of people.

When determining occupant loads with net factors, only consider the areas of the building that will actually be occupied. For example, if there is a room that is 30' × 30' (9.1 m × 9.1 m), and has four 12" × 12" (305 mm × 305 mm) columns, you have a gross square footage of 900 (83.61 m^2); however, if the chart calls for using net factors, then you would subtract the 4 square feet (38 cm^2) that the columns occupy and then end up with a figure of 896 net square feet (83.24 m^2). (page 120)

Challenging Question

This one-story freestanding structure has 11,250 ft^2 that will be used for exercise, plus equipment (150' × 75' [45 × 22.9 m]). The occupancy use is exercise room plus equipment. You note that for exercise room plus equipment, Table 1, (pages 120–121) shows a factor of 50, representing the square feet per person. Divide the factor into the area you are calculating, and you get an occupant load of 225 occupants.

Chapter 8: Fire Detection Systems

Matching

1. E (page 165)
2. H (page 165)
3. I (page 165)
4. C (page 166)
5. G (page 165)
6. B (page 165)
7. J (page 166)
8. A (page 165)
9. F (page 165)
10. D (page 165)

Multiple Choice

1. A (page 151)
2. D (page 151)
3. B (page 151)
4. C (page 152)
5. B (page 153)
6. A (page 153)
7. D (page 155)
8. C (page 155)
9. B (page 156)
10. A (page 157)
11. A (page 157)
12. D (page 157)
13. B (page 158)
14. C (page 158)

Fill in the Blank

1. Air sampling (page 158)
2. duct detectors (page 158)
3. activate the alarm (page 158)
4. temporal-3 pattern (page 159)
5. visual notification (page 159)
6. zoned system (page 159)
7. coded system (page 160)
8. noncoded (page 160)
9. Zoned noncoded (page 160)
10. zoned coded (page 160)

True/False

1. F (page 160)
2. T (page 160)
3. T (page 160)
4. F (page 161)
5. T (page 161)
6. F (page 161)
7. T (page 161)
8. T (page 162)
9. F (page 162)
10. F (page 162)

Labeling

1. An ionization smoke detector. (page 153)
2. Commercial ionization smoke detector. (page 156)
3. Beam detector. (page 157)
4. Flame detector. (page 158)
5. Air sampling detector air intake cone. (page 158)

Vocabulary

1. **Auxiliary system:** A fire alarm system that sounds an alarm in the building and transmits a signal to the fire department via a public alarm box system. (page 161)

2. **Bimetallic strip:** A device with components made from two distinct metals that respond differently to heat. When heated, the metals bend or change shape. (page 165)
3. **Ionization smoke detector:** A device containing a small amount of radioactive material that ionizes the air between two charged electrodes to sense the presence of smoke particles. (page 153)
4. **T tapping:** Improper wiring of an initiating device so that it is not supervised. (page 166)
5. **Temporal-3 pattern:** A standard fire alarm audible signal for alerting occupants of a building. (page 166)

Short Answer

1. **Ionization detector:**
   - Uses radioactive material within the device to detect invisible products of combustion.
   - Used to detect fires that do not produce large quantities of smoke in their early states. React quickly to fast-burning fires. Inappropriate for use near cooking appliances or showers.

   Photoelectric detector:
   - Uses a light beam to detect the presence of visible particles of smoke.
   - Used to detect fires that produce visible smoke. Reacts quickly to slow-burning smoldering fires. (page 153)
2. Most problems with smoke alarms are caused by a lack of power, dirt in the sensing chamber, or defective alarms.

   Power problems:
   - Smoke alarms require power from a battery or from a hard-wired 110-volt power source. Some smoke alarms that are hard-wired to a 110-volt power source are equipped with a battery that serves as a backup power source.
   - The biggest problem with battery-powered alarms is a dead or missing battery. Some people do not change batteries when recommended, resulting in inoperable smoke alarms. People may also remove smoke alarm batteries to use the battery for another purpose or to prevent nuisance alarm activations. The solution in this situation is to install a new battery in the alarm.
   - Hard-wired smoke detectors become inoperable if someone turns the power off at the circuit breaker. The solution in this situation is to turn the circuit breaker back on.
   - Most smoke alarms containing batteries signal a low-battery condition by emitting a chirp every few seconds. This chirp indicates that the battery needs to be replaced to keep the smoke alarm functional. Many hard-wired smoke alarms that are equipped with battery backups chirp to indicate that the backup battery is low, even when the 110-volt power source is operational. This feature assures that the smoke alarm will always have an operational backup system.

   Dirt problems:
   - A second problem with smoke alarms is the increased sensitivity that results when dust or an insect becomes lodged in the photoelectric or ionization chamber. Such an obstruction causes the alarm to activate when a small quantity of water vapor or smoke enters the chamber, leading to unnecessary alarms. The solution in this situation is to remove the cover of the chamber and gently vacuum out the chamber. Follow the manufacturer's instructions for this process.

   Alarm problems:
   - The third common problem you may encounter with smoke alarms is a worn-out detector. It is recommended that smoke alarms be replaced every 10 years because the sensitivity of the alarm can change. Some worn-out alarms may become overly sensitive and emit false alarms. Others may develop decreased sensitivity and fail to emit an alarm in the event of a fire. The date of manufacture should be stamped on each smoke alarm. If a detector is more than 10 years old, it should be replaced with a new alarm.
   - Understanding the basic functions and troubleshooting of simple smoke alarms enables you to respond to citizens who call your fire department when they encounter problems with their smoke alarms. Most importantly, it enables you to keep smoke alarms operational. Remember: Only *operational* smoke alarms can save lives. (page 156)
3. **Noncoded alarm:**
   - No information is given on what device was activated or where it is located.

Zoned noncoded alarm:

- Alarm system control panel indicates the alarm zone in the building that was the source of the alarm. It may also indicate the specific device that was activated.

Zoned coded alarm:

- The system indicates over the audible warning device which zone has been activated. This type of system is often used in hospitals, where it is not feasible to evacuate the entire facility.

Master-coded alarm:

- The system is zoned and coded. The audible warning devices are also used for other emergency-related functions. (page 160)

4. **Protected premises fire alarm system:** The fire alarm system sounds an alarm only in the building where it was activated. No signal is sent out of the building. Someone must call the fire department to respond.

   Remote supervising station system: The fire alarm system sounds an alarm in the building and transmits a signal to a remote location. The signal may go directly to the fire department or to another location where someone is responsible for calling the fire department.

   Auxiliary system: The fire alarm system sounds an alarm in the building and transmits a signal to the fire department via a public alarm box system.

   Proprietary supervising system: The fire alarm system sounds an alarm in the building and transmits a signal to a monitoring location owned and operated by the facility's owner. Depending on the nature of the alarm and arrangements with the local fire department, facility personnel may respond and investigate, or the alarm may be immediately retransmitted to the fire department. These facilities are monitored 24 hours a day.

   Central station system: The fire alarm system sounds an alarm in the building and transmits a signal to an off-premises alarm monitoring facility. The off-premises monitoring facility is then responsible for notifying the fire department to respond. (page 161)

Fire Alarm

1. When conducting the pre-occupancy inspection, any deficiencies need to be noted. There should be a copy of the results for you, the alarm company, and the owner. If you have to return, then only those items need to be inspected again. If the system meets your approval, then simply indicate on the paperwork that the system is approved. Often the alarm company has an inspection report that they use and will want you to sign. This paperwork should be kept in the occupancy file.

   Annual inspection paperwork should be submitted to the owner by an alarm company indicating the status of the system. If deficiencies are noted, continued follow-up must occur until the proper repairs a made. The most recent inspection paperwork needs to be kept on site at the control panel. Some alarm companies are installing tubes or boxes to house that paperwork. (page 163)

2. When a system is first installed, every part of the system needs to be tested. It is the only way you can sign off that all parts of the system worked when installed. On large systems this could take hours or days. Two alarm technicians with communication equipment should be present. One needs to be in the field actually testing the devices with you. The second needs to silence or reset the system and report what the control panel displays. The alarm company should perform an inspection themselves prior to your arrival to determine if everything works properly.

   The fire alarm and detection system should be designed to work on battery power, so for the test the 110-volt power should be shut off. This is the only way to tell if the audiovisual devices have enough power to operate properly. Each device component of the system needs to be activated to see that it functions as designed. If auxiliary functions such as releasing magnetically held doors or shutting down an HVAC unit are programmed, those must be checked as well to see that those functions occurred.

   If there is an interruption in the wiring, such as a missing detector, a trouble alarm should occur. The best way to check this is to actually take some devices off each circuit and see if a trouble alarm sounds.

   When inspecting audio visual devices, it should be noted that for those people who might be susceptible to seizures, the visual devices located within the space must flash together. They cannot each flash independently of each other.

   During the preoccupancy inspection, you should ensure that proper signals go to the monitoring station. This is also a time to verify that the proper alarm sequence was received by the monitoring system. Each of the three types of alarms—trouble, supervisory, and fire—are increasingly more important. A properly functioning alarm system has each type of alarm override the lower level alarm.

During the preoccupancy inspection, the circuit breaker needs to be identified and locked out at the breaker box to prevent an accidental shutting down of the fire alarm power. It is a good idea to also put the circuit number in the control panel for future use. The batteries in the control panel, and any power supplies in the field, must be marked with the year they were manufactured.

Once you are satisfied with the results of the final preoccupancy inspection, approval can be given. From this point on, unless the system is modified, you do not need to be this physically involved in the annual inspection process. (page 163)

Chapter 9: Fire Flow and Fire Suppression Systems

Matching

1. H (page 200)
2. J (page 198)
3. G (page 200)
4. C (page 199)
5. A (page 198)
6. I (page 200)
7. F (page 199)
8. E (page 198)
9. D (page 198)
10. B (page 198)

Multiple Choice

1. A (page 171)
2. A (page 172)
3. C (page 173)
4. A (page 174)
5. B (page 174)
6. D (page 176)
7. B (page 176)
8. C (page 177)
9. C (page 180)
10. A (page 180)
11. C (page 181)
12. D (page 181)
13. A (page 184)
14. B (page 184)
15. C (page 185)

Fill in the Blank

1. accelerators; exhausters (page 185)
2. preaction system (page 185)
3. deluge sprinkler system (page 185)
4. water (page 186)
5. dry chemical (page 187)
6. wet chemical (page 187)
7. sensitive electronic equipment (page 187)
8. Halon 1301 (page 188)
9. Carbon dioxide systems (page 188)
10. 40 psi; 24 hours (page 195)

True/False

1. F (page 189)
2. T (page 192)
3. T (page 189)
4. T (page 195)
5. F (page 189)
6. T (page 193)
7. F (page 189)
8. T (page 194)
9. F (page 191)
10. T (page 194)

Labeling

1. Upright sprinkler head. (page 189)
2. Pendent sprinkler head. (page 189)
3. Sidewall sprinkler head. (page 189)
4. Open sprinkler head. (page 190)
5. Nozzle sprinkler head. (page 190)
6. Flush sprinkler head. (page 190)
7. Recessed sprinkler head. (page 190)
8. Intermediate sprinkler head. (page 190)
9. Early-suppression fast-response (ESFR) sprinkler head. (page 191)
10. Fusible link sprinkler heads. (page 192)
11. Frangible sprinkler heads. (page 192)
12. Chemical-pellet sprinkler head. (page 192)

Vocabulary

1. **Accelerator:** A device that accelerates the removal of the air from a dry-pipe sprinkler system. (page 185)
2. **Distributors:** Relatively small-diameter underground pipes that deliver water to local users within a neighborhood. (page 174)
3. **Exhauster:** A device placed on the system side that accelerates the removal of the air from a dry-pipe sprinkler system. (page 185)
4. **Halon 1301:** A liquefied gas-extinguishing agent that puts out a fire by chemically interrupting the combustion reaction between fuel and oxygen. Halon agents leave no residue. (page 199)
5. **Riser:** The vertical supply pipes in a sprinkler system (NFPA 13). (page 199)

Short Answer

1. Most municipal water supply systems use both pumps and gravity to deliver water. Pumps may be used to deliver the water from the treatment plant to elevated water storage towers, which are aboveground water storage tanks designed to maintain pressure on a water distribution system or to reservoirs located on hills or high ground. These elevated storage facilities maintain the desired water pressure in the distribution system, ensuring that water can be delivered under pressure even if the pumps are not operating. When the elevated storage facilities need refilling, large supply pumps are used. Additional pumps may be installed to increase the pressure in particular areas, such as a booster pump that provides extra pressure for a hilltop neighborhood.

   A combination pump-and-gravity-feed system must maintain enough water in the elevated storage tanks and reservoirs to meet anticipated demands. If more water is being used than the pumps can supply or if the pumps are out of service, some systems are able to operate for several days by relying solely on their elevated storage reserves. Others may be able to function for only a few hours. (page 173)
2. NFPA 24, *Standard for the Installation of Private Fire Service Mains and Their Appurtenances*, recommends that fire hydrants be color coded to indicate the water flow available from each hydrant at 20 psi. It is recommended that the top bonnet and the hydrant caps be painted according to the following system:
   - Class C — Less than 500 gpm (1893 L/min) — Red
   - Class B — 500–999 gpm (1893–3784 L/min) — Orange
   - Class A — 1000–1499 gpm (3785–5677 L/min) — Green
   - Class AA — 1500 gpm and higher (5678 L/min) — Light blue

   (page 175)
3. **Static pressure** is the pressure in a water supply system when the water is not moving. Static pressure is potential energy because it would cause the water to move if there were some place the water could go. This kind of pressure causes the water to flow out of an opened fire hydrant. If there were no static pressure, nothing would happen when fire fighters opened a hydrant.

   Static pressure is generally created by **elevation pressure** and/or pump pressure. An elevated storage tank creates elevation pressure in the water mains. Gravity also creates elevation pressure (sometimes referred to as head pressure) in a water system as the water flows from a hilltop reservoir to the water mains in the valley below. Pumps create pressure by bringing the energy from an external source into the system.

   Static pressure in a water distribution system can be measured by placing a pressure gauge on a hydrant port and opening the hydrant valve. No water can be flowing out of the hydrant when static pressure is measured.

   When measured in this way, the static pressure reading assumes that there is no flow in the system. Of course, because municipal water systems deliver water to hundreds or thousands of users, there is almost always water flowing within the system. Thus, in most cases, a static pressure reading actually measures the normal operating pressure of the system.

   Residual pressure is the amount of pressure that remains in the system when water is flowing. When fire fighters open a hydrant and start to draw large quantities of water out of the system, some of the potential energy of still water is converted to the kinetic energy of moving water. However, not all of the potential energy turns into kinetic energy; some of it is used to overcome friction in the pipes. The pressure remaining while the water is flowing constitutes the residual pressure.

   Residual pressure is important because it provides the best indication of how much more water is available in the system. The more water flowing, the less residual pressure. In theory, when the maximum amount of water is flowing, the residual pressure is zero, and there is no more potential energy to push more water through the system. In reality, 20 psi is considered the minimum usable residual pressure, necessary to reduce the risk of damage to underground water mains or pumps.

Flow pressure measures the quantity of water flowing through an opening during a hydrant test. When a stream of water flows out through an opening (known as an orifice), all of the pressure is converted to kinetic energy. To calculate the volume of water flowing, measure the pressure at the center of the water stream as it passes through the opening, and then factor in the size and flow characteristics of the orifice. (page 176)

4. A Pitot gauge is used to measure flow pressure in psi (or kilopascals) and to calculate the flow in gallons (or liters) per minute.

   Obtain a Pitot pressure reading, by following these instructions:

   1. Remove the cap from a hydrant nozzle, preferably the 2½" (64 mm) discharge nozzle. Using the larger nozzle injects more air, and the reading must be slightly adjusted.
   2. Record the size of the discharge nozzle being used.
   3. Fully open the hydrant and allow water to flow.
   4. Hold the Pitot tube into the center of the flow and place it parallel to the discharge opening at a distance half of the inside diameter of the discharge nozzle.
   5. Record the pressure reading.
   6. Slowly close the hydrant valve fully and replace the cap.

   A Pitot pressure reading must be taken at each nozzle of the hydrant. When calculating the results, the flow results must be combined. For example, if the first nozzle flowed 750 gallons (2839 L) and a second nozzle flowed 825 gallons (3123 L), the total flow would be 1575 gallons (5962 L). (page 177)

5. The easiest way to determine the flow of a fire hydrant is to enter the static pressure, residual pressure, and Pitot pressure readings into a computer program. Within seconds the available fire flow at 20 psi (138 kPa) will be given. (page 178)

Fire Alarm

1. A building's use is an essential consideration in designing a sprinkler system that is adequate to protect against the hazards in a particular type of occupancy. NFPA 13 uses an occupancy approach to sprinkler system design. All buildings that fall within the scope of NFPA 13 fall into specific occupancy hazards. Each of the hazards has specific requirements for the spacing of sprinkler heads, the sprinkler head discharge densities, and water supply requirements. (page 179)
2. Dry chemical and wet chemical extinguishing systems are the most common specialized agent systems. Used in commercial kitchens, they protect the cooking areas and exhaust systems. In addition, some gas stations have dry chemical systems that protect the dispensing areas. These systems are also installed inside buildings to protect areas where flammable liquids are stored or used. Both dry chemical and wet chemical extinguishing systems are similar in basic design and arrangement.

   Dry chemical extinguishing systems use the same types of finely powdered agents as dry chemical fire extinguishers. The agent is kept in self-pressurized tanks or in tanks with an external cartridge of carbon dioxide or nitrogen that provides pressure when the system is activated.

   These five compounds are used as the primary dry chemical extinguishing agents:

   - Sodium bicarbonate—rated for class B and C fires only
   - Potassium bicarbonate—rated for class B and C fires only
   - Urea-based potassium bicarbonate—rated for class B and C fires only
   - Potassium chloride—rated for class B and C fires only
   - Ammonium phosphate—rated for class A, B, and C fires

   Wet chemical extinguishing systems discharge a proprietary liquid extinguishing agent. It is important to note that wet chemical extinguishing agents are not compatible with normal all-purpose dry chemical extinguishing agents. Only wet agents such as class K, or B:C-rated dry chemical extinguishing agents should be used where these systems are installed.

   All dry chemical extinguishing systems must meet the requirements of NFPA 17, *Standard for Dry Chemical Extinguishing Systems*. All wet chemical extinguishing systems must meet the requirements of NFPA 17A, *Standard for Wet Chemical Extinguishing Systems*. With both dry chemical and wet extinguishing agent systems, fusible-link or other automatic initiation devices are placed above the target hazard to activate the system. A manual discharge station is also provided so that workers can activate the system if they discover a fire. When the system is activated, the extinguishing agent flows out of all the nozzles. Nozzles are located over the target areas to discharge the agent directly onto a fire.

Many kitchen systems discharge agent into the ductwork above the exhaust hood as well as onto the cooking surface. This approach helps prevent a fire from igniting any grease buildup inside the ductwork and spreading throughout the system. Although the ductwork should be cleaned regularly, it is not unusual for a kitchen fire to extend into the exhaust system.

Dry and wet chemical extinguishing systems should be tied into the building's fire alarm system. Kitchen extinguishing systems should also shut down gas or electricity to the cooking appliances and exhaust fans.

Dry chemical extinguishing systems are installed at many self-service gasoline filling stations. (page 186)

Chapter 10: Portable Fire Extinguishers

Matching

1. E (page 218)
2. F (page 217)
3. H (page 217)
4. G (page 217)
5. I (page 218)
6. A (page 218)
7. J (page 218)
8. C (page 217)
9. D (page 217)
10. B (page 218)

Multiple Choice

1. B (page 204)
2. A (page 205)
3. D (page 206)
4. D (page 207)
5. B (page 207)
6. A (page 207)
7. B (page 210)
8. C (page 211)
9. C (page 212)
10. D (page 213)
11. A (page 213)
12. D (page 214)

Fill in the Blank

1. traditional (page 204)
2. type of fire (page 204)
3. characteristics; capabilities (page 207)
4. relative effectiveness (page 207)
5. prevent rekindling (page 207)
6. not been rated (page 208)
7. potential magnitude (page 208)
8. surface tension (page 210)
9. chemical chain reactions (page 210)
10. polar solvents (page 212)

True/False

1. T (page 209)
2. F (page 213)
3. T (page 214)
4. F (page 213)
5. F (page 214)
6. T (page 213)
7. T (page 212)
8. F (page 213)
9. F (page 206)
10. T (page 211)

Vocabulary

1. **Aqueous film-forming foam (AFFF):** A water-based extinguishing agent used on class B fires that forms a foam layer over the liquid and stops the production of flammable vapors. (page 212)
2. **Clean agent:** A volatile or gaseous fire extinguishing agent that does not leave a residue when it evaporates. Also known as a halogenated agent. (page 213)
3. **Polar solvent:** A water-soluble flammable liquid such as alcohol, acetone, ester, and ketone. (page 212)
4. **Saponification:** The process of converting the fatty acids in cooking oils or fats to soap or foam. (page 213)

Short Answer

1. Types of fires by class are:
    - Class A Ordinary combustibles
    - Class B Flammable or combustible liquids
    - Class C Energized electrical equipment
    - Class D Combustible metals
    - Class K Combustible cooking media (page 207)

2. The traditional lettering system uses the following labels:
   - Extinguishers suitable for use on Class A fires are identified by the letter "A" on a solid green triangle. The triangle has a graphic relationship to the letter "A."
   - Extinguishers suitable for use on Class B fires are identified by the letter "B" on a solid red square. Again, the shape of the letter mirrors the graphic shape of the box.
   - Extinguishers suitable for use on Class C fires are identified by the letter "C" on a solid blue circle, which also incorporates a graphic relationship between the letter "C" and the circle.
   - Extinguishers suitable for use on Class D fires are identified by the letter "D" on a solid yellow five-pointed star.
   - Extinguishers suitable for use on Class K (combustible cooking oil) fires are identified by a pictograph showing a fire in a frying pan. Because the Class K designation is new, there is no traditional-system alphabet graphic for it. (page 207)
3. Areas are divided into three risk classifications—light, ordinary, and extra hazard—based on the amount and type of combustibles that are present, including building materials, contents, decorations, and furniture.

   Light (low) hazard: Light (or low) hazard locations are areas where the majority of materials are noncombustible or arranged so that a fire is not likely to spread. Light hazard environments usually contain limited amounts of class A combustibles, such as wood, paper products, cloth, and similar materials. A light hazard environment might also contain some class B combustibles (flammable liquids and gases), such as copy machine chemicals or modest quantities of paints and solvents, but all class B materials must be kept in closed containers and stored safely. Examples of common light hazard environments are most offices, classrooms, churches, assembly halls, and hotel guest rooms.

   Ordinary (moderate) hazard: Ordinary (or moderate) hazard locations contain more class A and class B materials than light hazard locations. Typical examples of ordinary hazard locations include retail stores with on-site storage areas, light manufacturing facilities, auto showrooms, parking garages, research facilities, and workshops or service areas that support light hazard locations, such as hotel laundry rooms or restaurant kitchens.

   Ordinary hazard areas also include warehouses that contain class I and class II commodities. Class I commodities include noncombustible products stored on wooden pallets or in corrugated cartons that are shrink-wrapped or wrapped in paper. Class II commodities include noncombustible products stored in wooden crates or multilayered corrugated cartons.

   Extra (high) hazard: Extra (or high) hazard locations contain more class A combustibles and/or class B flammables than ordinary hazard locations. Typical examples of extra hazard areas include woodworking shops; service or repair facilities for cars, aircraft, or boats; and many kitchens and other cooking areas that have deep fryers, flammable liquids, or gases under pressure. In addition, areas used for manufacturing processes such as painting, dipping, or coating, and facilities used for storing or handling flammable liquids are classified as extra hazard environments. Warehouses containing products that do not meet the definitions of class I and class II commodities are also considered extra hazard locations. (page 209)
4. Portable fire extinguishers use seven basic types of extinguishing agents:
   - Water
   - Dry chemicals
   - Carbon dioxide
   - Foam
   - Wet chemicals
   - Halogenated agents
   - Dry powder (page 210)
5. Carbon dioxide also has several limitations and disadvantages:
   - Weight: Carbon dioxide extinguishers are heavier than similarly rated extinguishers that use other extinguishing agents.
   - The ratings on carbon dioxide extinguishers are far less than a typical multipurpose fire extinguisher.
   - Range: Carbon dioxide extinguishers have a short discharge range that requires the operator to be close to the fire, increasing the risk of personal injury.
   - Weather: Carbon dioxide does not perform well at temperatures below 0°F (−18°C) or in windy or drafty conditions because it dissipates before it reaches the fire.

- Confined spaces: When used in confined areas, carbon dioxide dilutes the oxygen in the air. If it is diluted enough, people in the space can begin to suffocate.
- Suitability: Carbon dioxide extinguishers are not suitable for use on fires involving pressurized fuel or on cooking grease fires. (page 212)

Fire Alarm

1. As a fire inspector, you assess the state of readiness by inspecting:
   - The travel distances to the fire extinguisher: The travel distance for a class A fire extinguisher is 75' (22.9 m). For a class B fire extinguisher, the proper travel distance is determined by the hazard being protected. NFPA 10 will dictate the distance and provides a table stating how many extinguisher are needed for a given floor area.
   - The mounting of the fire extinguisher: Most fire extinguishers are mounted with the top not higher that 5' (1.52 m) above the floor. If the fire extinguisher is more than 40 lbs (18.1 kg), then the top must be no higher than 3½' (1.07 m) above the floor. These distances make it easy for an operator to pick up the fire extinguisher. In either case, the bottom of the extinguisher must be a minimum of 4" (101.6 mm) off the floor. This allows for easy cleaning of the floor. The fire extinguisher must be mounted to prevent it from being moved to the back of a shelf or behind a piece of furniture. The fire extinguisher should be mounted near a normal means of egress and by an exit.
   - Access to the fire extinguisher: The fire extinguisher should be easily accessible. No item should be blocking the fire extinguisher from view or access.
   - The type of fire extinguisher: The proper type of fire extinguisher should be available. It would not be proper to have a water extinguisher in an electric room or a flammable liquids room. To prevent the chance of using the wrong fire extinguisher, most businesses use a multipurpose extinguisher. These extinguishers are good for the three common classes of fire: A, B, and C. An ABC-type and a BC-type extinguisher look identical, and they only way to tell them apart is to look at the rating or pictographs on the fire extinguisher.

   When you inspect fire extinguishers, look for physical damage to the container. Ensure that the hose is actually attached, no foreign matter is in the hose, the safety pin is in place, the seal holds the pin in place, and the gauge is showing the proper operating pressure. The location of the fire extinguisher should be considered as well. It should not be placed in an area where it will be subjected to possible damage. This would mean that it should not protrude into a means of egress more than 4" (101.6 mm). A fire extinguisher must have current inspection tags proving that an outside company inspected and tagged each unit. The tag is a visual indicator that the fire extinguisher was functional at the time of the outside company's inspection. (page 214)
2. It is common to find deficiencies during the course of a fire inspection. Examples include that the fire extinguisher has no pressure, the fire extinguisher was used, or the hose is not connected to the fire extinguisher. These items must be noted in the fire inspection report. In the inspection report, these items are easily addressed by writing that the fire extinguisher must be in proper working order with a new inspection tag. For issues such as fire extinguishers not being mounted, fire extinguishers that cannot be accessed, or fire extinguishers that are missing, write these as violations and reinspect for code compliance at a later date. A copy of the fire inspection report should be left with the building owner as a reference to correct the violations. (page 215)

Chapter 11: Electrical and HVAC Hazards

Matching

1. D (page 239)
2. E (page 238)
3. I (page 239)
4. C (page 238)
5. G (page 239)
6. J (page 239)
7. H (page 240)
8. A (page 238)
9. F (page 239)
10. B (page 238)

Multiple Choice

1. B (page 225)
2. D (page 224)
3. A (page 226)
4. B (page 225)
5. C (page 225)
6. C (page 227)
7. D (page 228)
8. C (page 228)
9. D (page 229)
10. D (page 230)
11. A (page 232)
12. D (page 233)

Fill in the Blank

1. Electrical inspectors (page 222)
2. volts (V) (page 224)
3. watts (W) (page 224)
4. fuse (page 224)
5. circuit breaker (page 225)
6. ground (page 225)
7. water-piping system (page 226)
8. local fire code (page 226)
9. transfer switch (page 226)
10. mechanical damage (page 228)

True/False

1. F (page 229)
2. F (page 231)
3. F (page 229)
4. T (page 233)
5. T (page 235)
6. T (page 230)
7. T (page 233)
8. F (page 236)
9. T (page 229)
10. T (page 230)

Vocabulary

1. **Amps:** The measure of the volume of electrical flow. (page 224)
2. **Conduits:** Round piping where wiring is routed through to provide protection from damage. (page 228)
3. **Gravity vents:** A component of a type of vent system for the removal of smoke from a fire that uses manually or automatically operated heat and smoke vents at roof level. The vents exhaust smoke from a reservoir bounded by exterior walls, interior walls, or draft curtains to achieve the design rate of smoke mass flow through the vents and include provision for makeup air (NFPA 204). (page 239)
4. **Plenum system:** An HVAC system that uses a compartment or chamber to which one or more air ducts are connected and that forms part of the air distribution system (NFPA 90A). (page 240)
5. **Raceway:** An enclosed channel of metal or nonmetallic materials designed expressly for holding wires, cables, or busbars, with additional functions as permitted in the electrical code. Raceways include, but are not limited to, rigid metal conduit, rigid nonmetallic conduit, intermediate metal conduit, liquid tight flexible conduit, flexible metallic tubing, flexible metal conduit, electrical nonmetallic tubing, electrical metallic tubing, underfloor raceways, cellular concrete floor raceways, cellular metal floor raceways, surface raceways, wireways, and busways (NFPA 70). (page 240)

Short Answer

1. Recognizing electrical problems is not complicated. If anything seems out of place, it is worthy of investigating. These are some warning signs of electrical problems:
   - A slight shock when handling appliances
   - Blinking lights or circuits that turn on and off by themselves
   - Insulation on wires that are broken or cracked
   - A computer monitor, television screen, or light that dims when a major appliance like an air conditioner turns on (page 222)
2. Electrical fires are caused principally by arcing or overheating. Arcing is a high-temperature luminous electric discharge across a gap. Overheating is generally the result of excess current, excessive insulation, or poor connections.
   - Gaps may be created in the normal operation of equipment, such as in switches or in motors with brushes. They may also be created at loose splices or terminals or where the insulation around the wiring has broken down or has been damaged and the wire is in close proximity to another wire or a grounded metal surface. The arc is commonly referred to as a "short" or "short circuit" because the electricity takes a shorter path rather than going through the full circuit. Arcing produces enough heat to ignite nearby combustible materials, such as insulation, and can throw off particles of hot metal that can cause ignition. Arcing can also melt wiring and produce sparks.
   - Overheating is more subtle, harder to detect, and slower to cause ignition but is equally capable of causing a fire. Conductors and other electrical equipment may generate a dangerous level of heat when they carry excess current. This excess of current may cause wiring to overheat to the point at which the temperature is sufficient to ignite nearby combustible materials, such as wood framing. For example, a lightweight extension cord is run under a rug to power a space heater that uses more electricity than the extension cord is designed to handle. The extension cord overheats and the rug catches on fire. (page 224)
3. Electrical systems have overcurrent protection that opens a circuit if the amount of current will cause an excessive or dangerous temperature. Fuses and circuit breakers are the most commonly used overcurrent devices.

A fuse is an overcurrent protective device with a circuit-opening fusible part that is heated and severed by the passage of overcurrent through it. It consists of a thin wire surrounded by a casing. If a wiring short occurs, the wire in the fuse will overheat very quickly and fail, breaking the electrical current before a fire can break out at the point of the arc. Once the fuse wire breaks, the fuse must be replaced. Fuses are rated by their amperage protection. Fuses are located in a central fuse box.

There are two general categories of fuses: plug and cartridge. Plug fuses come in a wide variety of types and styles, which can range from quick acting to time delay and from a standard base to a tamperproof base. Plug fuse boxes have round sockets that look much like a light bulb socket. The fuse looks like a plug that screw into the socket.

The other type of fuse is the cartridge fuse. Like plug fuses, cartridge fuses are available in quick-acting and time-delay types. Some are designed for onetime use and are thrown away after they trip. The other type of cartridge fuse is the renewable link cartridge fuse where the internal components can be replaced by an electrician and then reinstalled. The renewable link cartridge fuses can fail to operate properly if not properly serviced, so they should only be serviced by qualified persons, such as experienced electricians or the cartridge fuse manufacturer's technicians.

Another form of overcurrent protection device is the circuit breaker. The circuit breaker operates based on a variety of principles, but all have the same result: A switch is opened to stop the flow of electricity through the circuit. Because there is a switch rather than a wire that breaks, the circuit breaker can be reset once the fault condition is resolved. Frequently circuit breakers use more than one type of technology to activate the most effectively to the specific fault condition.

Some circuit breakers may have a shunt trip that allows them to be operated from remote locations. For example, a remote circuit breaker may be used to shut down equipment under kitchen exhaust hoods in restaurants or the main disconnect on the incoming utility service feed to a building.

Additional specialized circuit breakers include ground-fault circuit interrupter (GFCI) protection and arc-fault circuit interrupter (AFCI) protection. Any of these devices may feel warm under normal loads, but none should be too hot to touch. (page 224)

4. As a fire inspector, you need to be on the lookout for the common problems associated with wiring. These problems often stem from either improper installation or damage to existing wiring. These conditions may allow electrical wiring to short, arc, or become overloaded and can become an ignition source, especially in hazardous locations. Be on the lookout for damaged insulation, broken wires, and sloppy repairs to electrical wiring. Wiring must be supported properly along its length and at the point at which it terminates. Wiring should not be exposed to excessive external heat. (page 228)

5. Several methods for managing the movement of smoke in a building may involve HVAC systems in concert with the building's construction. Some design concepts rely on natural ventilation and compartmentalization; others actively use portions of an HVAC system to either contain or remove smoke.

Natural ventilation through the use of gravity vents is most commonly found in some large warehouses. This application of smoke management relies on the buoyancy of the hot products of combustion to open a fusible-link operated vent in a high ceiling. In a few warehouse applications, the gravity vents are replaced by exhaust fans on the roof that automatically activate in the event of a fire. Protection of the fan drive motor, bearings, and electrical supply is required for these fans to operate in high-heat conditions.

Older high-rise designs also utilize manually opened vents or operable or breakable windows so fire fighters can manually provide an avenue for fresh air to enter and smoke to leave an upper story of a building. With this setup, the building's HVAC system usually shuts down the air-handling units, if so equipped.

Compartmentation, also referred to as passive smoke management, relies on interior smoke barrier construction in the walls and floor/ceiling assemblies of a building to restrict the free movement of smoke from one smoke compartment to another. In this application, the HVAC system is sent a command from the fire alarm system to shut down the air-handling units and close all supply-and-return air dampers in and surrounding the affected smoke compartment. The size and location of the smoke compartments in a building is determined by the building's designers. When inspecting a building using this type of smoke management design, you need to know where smoke barrier walls have been constructed and must verify during subsequent inspections that the integrity of the barrier walls have not been compromised. Initial acceptance testing and annual testing by a qualified person, such as a fire protection engineer, mechanical engineer, or an air-balance technician, is also necessary to ensure that the HVAC control systems, air-handling unit shutdown features, and smoke/fire dampers operate as designed.

A third smoke management method is intended to contain smoke to its zone of origin by combining compartmentation and using all or a portion of the HVAC system to depressurize the smoke zone. Upon activation, air-handling units or dedicated smoke management fans exhaust air from the smoke zone.

Smoke/fire dampers close, supply air is shut down, and a negative pressure is developed in the smoke zone where the fire is located. Referred to in the model codes as the pressurization method, its name is a misnomer because the system must actively depressurize the zone; otherwise expansion forces from the fire would force smoke through the smoke barriers. With this method, any leakage in the construction of the smoke barriers will cause air to leak into the zone of origin, meaning the portion of the duct system including the smoke zone, instead of having smoke leak out of the zone. Smoke is not actually removed in sufficient quantities to keep the smoke zone habitable. Introducing outside air to flush out the smoke is not recommended: The newly introduced air (called makeup air) may be contaminated, the balance of outside air and exhaust air is subject to weather conditions and seasonal changes, and the reliability of the system is much harder to maintain. Detailed information on this design concept and factors to consider in the design, installation, commissioning, and maintenance of this type of smoke management system can be found in NFPA 92A, *Standard for Smoke-Control Systems*. Due to its complexity, the design, commissioning, acceptance testing, and annual operational testing of this type of smoke management system should be performed by qualified and experienced engineering, air balance, and technical personnel.

The fourth smoke management system can be found in NFPA 92B, *Standard for Smoke Management Systems in Malls, Atria, and Large Areas*. This design concept is used in areas with high ceilings and employs HVAC equipment, either dedicated to the smoke management use or part of the normal HVAC system, to remove products of combustion accumulated above a calculated height above the floor. Makeup air, carefully controlled so as not to distort the fire plume and resulting exhaust calculations, is also an important part of this design methodology. It is the intent of this design concept to maintain a tenable environment for a minimum amount of time necessary to evacuate occupants from the building. Again, due to the complexities involved, only qualified and experienced personnel, such as fire protection engineers and mechanical engineers, should be involved in the engineering and testing of this type of smoke management system. (page 234)

Fire Alarm

1. Transformers are may be found in large industrial facilities that require large amounts of voltage to run mechanical processes. Dry-type transformers and fluid-filled electrical transformers are used in industrial and commercial occupancies. Dry-type transformers use air as a temperature-maintenance mechanism; fluid-filled transformers typically use a type of oil to remain cool. Dry-type transformers are common in newer commercial facilities with low to moderate current demands. Fluid-filled transformers are more common in industrial facilities and as the main transformer on incoming services where large current loads are present due to their more efficient operation.

   In most cases, dry-type transformers do not require a separate room or vault, but they must be separated from combustible materials and provided with adequate ventilation to remove excess heat. Fluid-filled transformers, if not installed outdoors in a fenced yard or substation, usually are required to be installed in a vault with 3-hour fire-resistive construction and associated opening protection. The vault must also be curbed or of sunken construction to contain the contents of the transformer and any fire protection system water discharge should a spill occur.

   Older fluid-filled transformers often used mineral oil as their cooling medium. Newer transformer fluids, classified as less flammable or nonflammable, are now available in addition to mineral oil. When these new fluids are used, the requirements for vaults are reduced or eliminated. Flammable oils may still be found in new installations, so you should inquire as to the type and quantity of material used. Polychlorobiphenyl (PCB) fluid, although widely used in the past, is no longer available due to environmental concerns. These transformers are required to be prominently marked with their contents and eventually will be phased out of use.

   Under conditions of a full electrical load, transformers operate at elevated temperatures. Many will be too hot to touch for more than a few seconds. All transformers should be provided with adequate ventilation, and the clearance requirements marked on the transformer should be maintained. Materials should not be stored on top of any transformer enclosure.

   Outdoor transformers should be located in such a way that leaking fluids will drain away from buildings and prevented from entering environmentally sensitive areas. These transformers should be positioned in such a manner that they will not expose building exits or windows to fire in the event of a transformer failure. Transformers should also be protected from each other to avoid the failure of one transformer from causing the failure of an adjacent one. Protection may involve the construction of a freestanding firewall between transformers, the installation of a water-spray deluge system, or a combination of the two. The size and criticality of service provided by a transformer may dictate additional protection features due to the extended time period to receive a replacement for a damaged or destroyed transformer. (page 227)

2. Design and testing of smoke management systems requires careful and diligent work by qualified and experienced professional designers. Information and guidance can be found in a number of publications, including the NFPA/SFPE *Fire Protection Engineering Handbook*. Initial testing at the time of construction and recurrent testing at intervals established by either the initial designers or in accordance with NFPA standards should be conducted only by experienced and qualified fire protection engineers, mechanical engineers, and air balance technicians.

   Commissioning, acceptance testing, and related documentation of these systems are especially important because they can be very complex. Reliability, likewise, depends on the complexity of the design and the maintenance provided during the life of the facility. Annual testing by qualified professional engineers and air balance technicians of these systems is greatly facilitated by good original commissioning documentation. (page 235)

Chapter 12: Ensuring Proper Storage and Handling Practices

Matching

1. I (page 252)
2. C (page 247)
3. E (page 259)
4. J (page 252)
5. H (page 246)
6. F (page 248)
7. B (page 260)
8. A (page 247)
9. D (page 262)
10. G (page 257)

Multiple Choice

1. B (page 245)
2. A (page 252)
3. A (page 252)
4. B (page 252)
5. D (page 253)
6. A (page 251)
7. A (page 251)
8. A (page 251)
9. A (page 251)
10. C (page 250)
11. C (page 246)
12. B (page 246)
13. D (page 264)
14. A (page 262)
15. C (page 259)

Fill in the Blank

1. NFPA 30, *Flammable & Combustible Liquids Code* (page 245)
2. combustible (page 246)
3. poison gas (page 247)
4. compressed gas (page 247)
5. liquefied gas (page 247)
6. hazardous material (page 249)
7. being transported (page 249)
8. NFPA 704 (page 251)
9. flammable and combustible liquids (page 254)
10. 120 gallons (454 L) (page 255)

True/False

1. T (page 255)
2. F (page 256)
3. F (page 257)
4. F (page 258)
5. T (page 255)
6. T (page 257)
7. T (page 260)
8. F (page 255)
9. F (page 261)
10. T (page 257)

Vocabulary

1. **Bills of lading:** Shipping papers for roads and highways. (page 252)
2. **Cryogenic gas:** A refrigerated liquid gas having a boiling point below –130°F (–90°C) at atmospheric pressure (NFPA 1992). (page 247)
3. **Dewar containers:** Containers designed to preserve the temperature of the cold liquid held inside. (page 258)
4. **Safety Data Sheet (SDS):** A form, provided by manufacturers and compounders (blenders) of chemicals, containing information about chemical composition, physical and chemical properties, health and safety hazards, emergency response, and waste disposal of the material. (page 252)
5. **Pyrophoric gases:** Flammable gases that spontaneously ignite in air; examples of pyrophoric gases include silane and phosphine. (page 246)

Short Answer

1. NFPA 55, *Compressed Gases and Cryogenic Fluids Code*, categorizes gases into two classifications: Physical and Health.
2. These are the nine DOT chemical families recognized in the ERG:
   - DOT class 1, Explosives
   - DOT class 2, Gases
   - DOT class 3, Flammable combustible liquids
   - DOT class 4, Flammable solids
   - DOT class 5, Oxidizers
   - DOT class 6, Poisons
   - DOT class 7, Radioactive materials
   - DOT class 8, Corrosives
   - DOT class 9, Other regulated materials (ORMs) (page 250)
3. The ERG is divided into four colored sections: yellow, blue, orange, and green.
   - **Yellow section:** Chemicals in this section are listed numerically by their four-digit UN or NA identification number. Entry number 1017, for example, identifies chlorine. Use the yellow section when the UN number is known or can be identified. The entries include the name of the chemical and the emergency action guide number.
   - **Blue section:** Chemicals in this section are listed alphabetically by name. The entry includes the emergency action guide number and the identification number. The same information, organized differently, is in both the blue and yellow sections.
   - **Orange section:** This section contains the emergency action guides. Guide numbers are organized by general hazard class and indicate what basic emergency actions should be taken, based on hazard class.
   - **Green section:** This section is organized numerically by UN and NA identification number and provides the initial isolation distances for specific materials. Chemicals included in this section are highlighted in the blue or yellow sections. Any materials listed in the green section are always extremely hazardous. (page 250)
4. Generally, an SDS includes:
   - Physical and chemical characteristics
   - Physical hazards of the material
   - Health hazards of the material
   - Signs and symptoms of exposure
   - Routes of entry
   - Permissible exposure limits
   - Responsible party contact
   - Precautions for safe handling (including hygiene practices, protective measures, and procedures for cleaning up spills or leaks)
   - Applicable control measures, including personal protective equipment
   - Emergency and first aid procedures
   - Appropriate waste disposal (page 252)
5. Pesticide bags must be labeled with specific information. Fire fighters can learn a great deal from the label, including:
   - Name of the product
   - Statement of ingredients
   - Total amount of product in the container
   - Manufacturer's name and address

- U.S. Environmental Protection Agency (EPA) registration number, which provides proof that the product was registered with the EPA
- The EPA establishment number, which shows where the product was manufactured
- Signal words to indicate the relative toxicity of the material:
  - Danger: Poison: Highly toxic by all routes of entry
  - Danger: Severe eye damage or skin irritation
  - Warning: Moderately toxic
  - Caution: Minor toxicity and minor eye damage or skin irritation
- Practical first aid treatment description
- Directions for use
- Agricultural use requirements
- Precautionary statements such as mixing directions or potential environmental hazards
- Storage and disposal information
- Classification statement on who may use the product
  In addition, every pesticide label must have the statement "Keep out of reach of children." (page 260)

Labeling

1. An MC-306 flammable liquid tanker. (page 263)
2. An MC-307 chemical hauler. (page 263)
3. An MC-312 corrosives tanker. (page 263)
4. An MC-331 pressure-cargo tanker. (page 263)
5. E. An MC-338 cryogenic tanker. (page 263)
6. A tube trailer. (page 263)
7. A dry bulk cargo tank. (page 263)
8. Intermodal tanks. (page 263)

Fire Alarm

1. In regions with large chemical and petrochemical plants, pipelines may also carry a variety of intermediate chemicals used in the production of finished products such as plastic resins and refined petroleum products. Pipelines are often buried underground, but they may be above ground in remote areas. The pipeline right-of-way is an area, patch, or roadway that extends a certain number of feet on either side of the pipe itself. This area is maintained by the company that owns the pipeline. The company is also responsible for placing warning signs at regular intervals along the length of the pipeline.

   Pipeline warning signs include a warning symbol, the pipeline owner's name, and an emergency contact phone number (page 264)
2. You can assist emergency response personnel in your jurisdiction by documenting and forwarding emergency contact, facility site plans, emergency shutdown, and facility access information to the fire department. Prelocated caches of response equipment, such as foam concentrate, spare self-contained breathing apparatus (SCBA) air bottles, and fire hose, can be established in cooperation with the pipeline owner at key locations in their facilities. You can be a vital link in communications between the pipeline owner and the fire fighters responding to an incident. (page 264)
3. Fire protection systems for hazardous materials vary greatly depending on the chemical composition of the hazard. Differing hazardous materials may have very different reactions to fire extinguishing agents. Highly reactive metal hydrides react violently with water, for example.

   Containment of hazardous materials are a concern in the design of any fire extinguishing system. The runoff caused by the leak from a tank, container, or piping system may be aggravated and multiplied many times by the application of large hose streams to a fire. Impounding contaminated runoff from a facility using or warehousing hazardous materials should be considered in the planning stage of any project.

   Foam-water automatic sprinkler systems and high- and low-expansion foam systems, if they are chemically compatible with the hazardous material, can provide not only effective fire suppression capability but also can suppress potentially hazardous or toxic vapors.

Gaseous agents, both local application and total flooding systems, may be used to protect hazardous materials. A wide range of gaseous agents are available, including carbon dioxide, halocarbon clean agents, and inert clean agents. Each agent has its advantages and its drawbacks. NFPA 2001, *Standard on Fire Clean Agent Extinguishing Systems*, should be consulted for design requirements.

Due to the complexities involved in the design of a fire protection system for hazardous materials, a fire protection engineer should be consulted to review the hazards present and the performance of any proposed design. (page 266)

Chapter 13: Safe Housekeeping Practices

Matching

1. I (page 286)
2. J (page 286)
3. H (page 286)
4. F (page 286)
5. G (page 286)
6. D (page 286)
7. A (page 286)
8. E (page 286)
9. B (page 286)
10. C (page 286)

Multiple Choice

1. B (page 275)
2. C (page 275–276)
3. A (page 275)
4. A (page 277)
5. B (page 276)
6. D (page 277)
7. D (page 277)
8. C (page 278)
9. B (page 279)
10. B (page 279)

Fill in the Blank

1. egress system (page 279)
2. fire protection (page 279)
3. effective housekeeping practice (page 279)
4. dunnage (page 280)
5. emergency management plan (page 280)
6. disposal plans (page 280)
7. 18" (45.7 cm) (page 281)
8. spray booth (page 280)
9. fire safety risk (page 281)
10. Smoking (page 282)

True/False

1. T (page 283)
2. T (page 281)
3. T (page 280)
4. F (page 282)
5. F (page 281)
6. T (page 279)
7. F (page 281)
8. T (page 279)
9. T (page 280)
10. F (page 281)

Vocabulary

1. **Aspect:** Compass direction toward which a slope faces (NFPA 1144). (page 277)
2. **Dunnage:** Loose packing material (usually wood) protecting a ship's cargo from damage or movement during transport (NFPA 1405). (page 280)
3. **Fuel ladder:** A continuous progression of fuels that allows fire to move from brush to limbs to tree crowns or structures. (page 277)

Short Answer

1. Safe housekeeping practices—both indoor and outdoor—accomplish four major fire and life safety objectives:
   - Eliminate unwanted fuels, helping to control fire growth and making extinguishment easier.
   - Remove obstructions or impediments to egress.
   - Control sources of ignition.
   - Improve safety for firefighting and emergency response personnel. (page 274)

2. Poor housekeeping practices outside of the building can result in:
   - Obstructions to the site or building
   - Obstructions to fire protection equipment
   - Fire exposure threats
   - Wildfire concerns
   - An unattractive nuisance easily ignited by vandals or juveniles. (page 275–276)
3. Three basic requirements of good housekeeping are:
   - Equipment arrangement and layout, which includes properly cleaning, servicing, and placing devices that generate combustible dusts or waste materials. Sometimes the layout or location of equipment can pose additional problems, such as equipment that is producing dust located close to an ignition source.
   - Material storage and handling, which includes trash and waste disposal or recycling operations that are found in buildings or on the property outside of a building.
   - Operational neatness, cleanliness, and orderliness, which includes emptying trash and waste on frequent intervals to avoid accumulation. The frequency of trash removal may vary substantially depending on the type, form, and amount of trash or waste generated. Some occupancies, such as large retail stores, have very complex and involved trash and waste-handling operations to remove, compact, and bale materials for recycling or disposal. (page 275)
4. The fuel ladder is a continuous progression of fuels that allows fire to move from brush to limbs to tree crowns or structures. Depending on a number of factors, including slope, meaning the upward or downward slant of the land; aspect, the compass direction toward which a slope faces; and environmental factors such as weather and winds conditions expected during fire events, the defensible space may need to vary from as little as 30' (9 m) to as much as 200' (61 m). It is not necessary to remove all vegetation in this space: modifying existing vegetation and fire-safe landscaping treatments appropriate for the climate can achieve a practical solution. Where conditions cannot provide the necessary defensible space, relocating or removing outdoor storage or protecting the structure through improved construction materials, especially roofing, can provide the necessary protection. (page 277)
5. Processes that generate lint, especially clothes-drying operations, need to be cleaned frequently to minimize buildup of this easily ignitable material. This can especially be a problem with commercial laundry operations that handle large volumes and have dryers heated to higher temperatures to dry clothes more quickly. You should look for evidence of lint accumulation inside the laundry equipment, exhaust ductwork, and in the room itself.

   Timber, woodworking, textile or agricultural grain processing facilities can generate large amounts of combustible dusts or fibers that, when airborne, can explosively ignite or rapidly spread a fire faster than a conventional sprinkler system can control. Dust and fibers can accumulate on walls, ceilings, motors, heating equipment, and structural members; under tables; and inside ducts and conveying equipment. Removing these dusts or fibers can be a dangerous process because an explosive fire can occur if not done correctly. In most cases dust, lint, and fibers should be removed by way of a vacuum system using dust-collection equipment and safe electrical hardware. In a few cases, damp cloths are used to remove dust or fibers, and then they are properly disposed of in metal containers if spontaneous combustion is a possibility. Compressed air or blowers should never be used to clean dusty areas because this only suspends the material in the air, making it easy to ignite. (page 278)

Fire Alarm

1. Many storage and retail occupancies, sometimes referred to as big-box stores, use high-piled storage arrangements to maximize space. High-piled storage is often defined as solid-piled, palletized, rack storage, bin box, or shelf storage in excess of 12' (366 cm) in height. In some cases the storage arrangement poses risks to firefighting personnel by creating very narrow aisles and unstable piles. This arrangement could hamper firefighting operations, and the piles could collapse on fire fighters, especially when boxes became saturated with water or weakened by fire damage.

   NFPA 1 contains specific minimum aisle dimensions and maximum pile height and sizes. The minimum aisle dimensions and storage pile dimensions are based on a number of factors, including the commodity being stored and level of fire protection provided. Aisles between racks of storage or solid piles on pallets should be a minimum of 4' (1.2 m) according to NFPA 13, *Standard for the Installation of Sprinkler Systems*. This distance is needed for safe forklift operations inside the building. In some rack storage conditions, NFPA 13 requires aisles to be a minimum of 8' (2.4 m).

   Another concern that occurs in high-piled storage operations involves safety concerns for products that expand with the absorption of water. Where large quantities of paper products, especially rolled paper, are stored, it is necessary to keep an aisle between the paper storage and the sidewall of the building. During firefighting operations, either from automatic fire sprinklers or firefighting hoselines, these products can soak up water, expand, and push against the outside walls of the building, causing severe structural damage or collapse. For this reason, NFPA 13 requires a 24" (610 mm) aisle between exterior walls of the building and materials that will absorb water and expand. (page 279)

2. Improperly stored materials can obstruct access to fire protection equipment, such as fire extinguishers, control valves, and fire alarm pull-stations. Improper storage can also impair the proper operation of passive fire protection equipment, such as fire doors.

   In sprinkler-protected buildings, NFPA 1 and NFPA 13 require that storage be kept at least 18" (4.57 cm) below sprinklers. This distance allows sprinklers to develop their characteristic umbrella spray pattern to effectively extinguish a fire. Storage too close can hamper or impair ceiling-mounted sprinklers. (page 281)

Chapter 14: Writing Reports and Keeping Records

Matching

1. D (page 300)
2. E (page 300)
3. A (page 300)
4. C (page 300)
5. B (page 304)
6. H (page 299)
7. I (page 299)
8. F (page 301)
9. G (page 299)
10. J (page 294)

Multiple Choice

1. D (page 300)
2. C (page 294)
3. D (page 291)
4. D (page 293)
5. C (page 294)

Fill in the Blank

1. complaint (page 296)
2. reinspections (page 299)
3. Field notes (page 291)
4. written record (page 291)
5. diagrams (page 293)
6. Letters (page 294)
7. final notice (page 294)
8. e-mails (page 296)
9. Photographs (page 293)
10. legal proceeding (page 302)

True/False

1. F (page 299)
2. T (page 296)
3. F (page 293)
4. F (page 291)
5. T (page 290)
6. F (page 294)
7. T (page 299)
8. T (page 299)
9. T (page 291)
10. T (page 293)

Short Answer

1. Each letter written should include these basic design elements:

Header	Agency letterhead or return address
	Date
	Inside address
	Subject/Reference
Salutation	Dear Mr., Ms., or other recognized and appropriate form
Body	Start with an introduction, followed by the main topic(s) of the letter and ending with a closing paragraph or statement. Start a new paragraph for every major point.
Closing/Signature	Begin with "Sincerely" or "Regards," and end with your signature, printed name, and title.
Enclosures	List all documents or other information included with your letter.
Copies	List the name of everyone who received a copy of this letter, followed by their company or agency name. (page 296)

2. Contact information includes the following information for each property owner and tenant:
   - Company name and address
   - Primary point of contact information, including telephone number, fax number, Web site, and e-mail address (page 294)
3. Property information includes:
   - Location
   - Occupancy type and primary use
   - Construction type
   - Significant fire protection features, such as fire detection and suppression systems (page 294)
4. All fire inspection reports should contain these common elements:
   - Be professional and well written.
   - Be focused and concise.
   - Present facts, free from bias, opinion, or criticism.
   - Use correct grammar, word usage, spelling, punctuation, and style. Computer-generated reports and templates can ensure consistent appearance, format, and quality.
   - Whenever possible, avoid passive sentences. Use "The property owner shall correct the following" instead of "The following shall be corrected." (page 293)
5. Common uses of written records include:
   - Plans review and preconstruction design meetings
   - Fire inspections, both initial and follow-up
   - Appeal hearings
   - Legal proceedings
   - Other uses, such as fire investigations, staff scheduling and job assignments, training and education, trend analysis, media reports, development of policies and procedures, and budget resource justification (page 290)

Fire Alarm

To be reviewed by your instructor.

Chapter 15: Life Safety and the Fire Inspector

Matching

1. E (page 314)
2. I (page 314)
3. J (page 314)
4. H (page 314)
5. A (page 314)
6. D (page 314)
7. B (page 314)
8. F (page 314)
9. C (page 314)
10. G (page 314)

Multiple Choice

1. D (page 308)
2. B (page 309)
3. A (page 309)
4. D (page 311)
5. A (page 308)
6. D (page 312)
7. C (page 308)
8. D (page 310)

Fill in the Blank

1. public information (page 309)
2. effectiveness metrics (page 313)
3. Direct costs (page 312)
4. frequency; severity (page 310)
5. stakeholder groups (page 311)
6. Indirect costs (page 312)
7. fire safety (page 308)
8. anecdotal evidence (page 309)
9. National Fire Incident Reporting System (NFIRS) (page 310)
10. local (page 310)

True/False

1. T (page 312)
2. T (page 311)
3. T (page 312)
4. F (page 312)
5. F (page 310)
6. F (page 309)
7. T (page 312)
8. T (page 308)
9. T (page 310)
10. F (page 313)

Vocabulary

1. **Direct costs:** Expenses that must be paid but would otherwise not have occurred without the program such as printing, supplies, or fuel. (page 314)
2. **Fire and life safety educators:** Personnel who teach messages about fire safety to target audiences. (page 308)
3. **National Fire Incident Reporting System (NFIRS):** A national database that collects data about fire incidents including the collection of many of the underlying factors that caused the fire. (page 310)
4. **Public relations:** Activities that are focused on building a positive image of any organization. (page 308)

Short Answer

1. For the fire and life safety education program to be successful, it must be well thought out with goals established. Some of the questions to consider include:
   - What do you want to achieve?
   - Do you want to educate every student in a certain age group?
   - Will you provide refresher education and training as this group matures?
   - Are you going to go to the students or have them come to you?
   - Are you going to use videos or distribute gifts with a fire safety message?
   - Who will present the educational programs?
   - Who will train the educators?
   - How will the program be financed?
   - How long will the program run?
   - When will it start?

   These questions can only be answered once the goals and objectives have been established. (page 311)
2. To measure the effectiveness of a fire and life safety program, first measure the behaviors driving the problem before the fire and life safety public education program begins. Then examine local data to see if there is a change in behavior. The results should be compared to the goals and objectives of the program. If it is not successful, it should be modified or abandoned in favor of other programs that can show their benefits. For example, after a series of fire deaths, you might assess the percentage of homes without a working smoke alarm either due to nonexistence or a dead or missing battery. You would then present your fire and life safety educational program to the community. After the presentation, you would reassess the percentage of homes without a working smoke alarm. Any positive change should be attributable to your program. More smoke alarms *should* translate into fewer fire deaths. (page 313)
3. Using an example of fires started by unattended matches and lighters, city council members are stakeholders because the public looks to them to resolve community problems. The hospital's burn unit is a stakeholder because they have to treat the people who are burned as a result of a fire. The insurance companies are stakeholders because they pay the claims for the property damage as well as the heath care. Neighborhood groups are stakeholders because high fire incidents in the neighborhood reduce property values. Do not forget to include the fire department as a stakeholder. (page 311)
4. As a fire inspector, you must have the ability to develop programs that will address the risks of fire. Your programs should include the following objectives:
   - Target behaviors that contribute to the likelihood of fires occurring
   - Work to reduce injuries and damage if a fire does occur (page 308)

Fire Alarm

Once data is collected, you will quickly see that some events occur more often (*frequency*) than others and that some of the problems are more significant when they do occur (*severity*) than others. To maximize available resources, create a matrix plotting the problems based on frequency and severity. The most common method is to plot the frequency of a specific risk on a horizontal axis. It is recommended to use a rating scale, such as 1 to 10, to give the relative degree of frequency. Then plot each problem by its severity on the vertical axis, again using a 1 to 10 scale. The goal of creating the matrix is to focus the public education message on those issues with the greatest risk to the community.

Once the problems have been plotted, you will find some problems arise more frequently and have a high degree of severity. These types of problems will be in the upper right corner of the graph. Those with low frequency and low severity will be in the lower left corner of the graph. This allows you to quickly determine the highest priority of problems to address. For example, a community in Kansas might have tornados as a high-frequency, high-severity event and categorize an earthquake as a low-frequency, high-severity event because they are not near a fault line. This might highly contrast with a community along the California coast, where tornados would be very low frequency whereas earthquakes would have a much higher frequency. (page 310)

Fire Inspector Level II Activities

Review the photographs below and answer the associated questions accordingly.

1. 

While conducting an acceptance test for a new sprinkler system installation, you observe this condition. Please describe the fire code violation.

____

____

____

2. 

On a routine inspection of a residential occupancy, you notice the smoke alarm shown. Identify any fire safety concerns with this smoke alarm.

____

____

____

3. 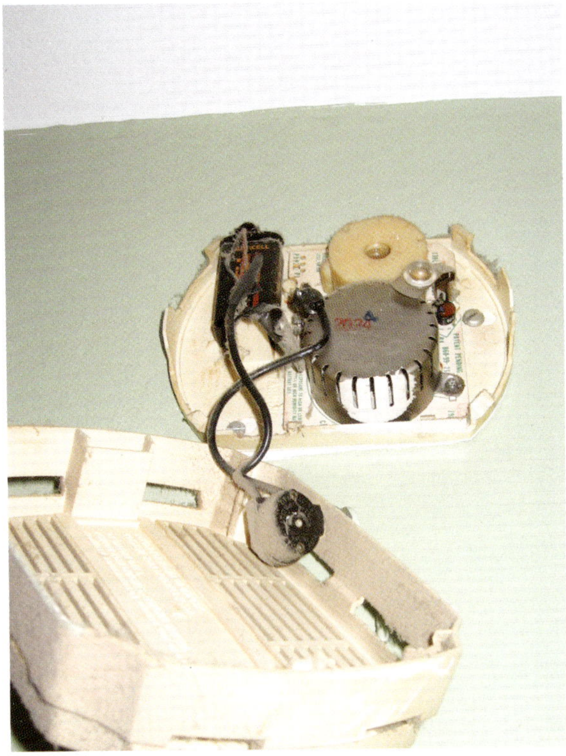

A closer examination of the residential smoke alarm in Question 2 reveals a possible fire safety concern. Identify this concern.

4. 

On a routine inspection you observe this furnace room. What fire safety concerns do you have?

5. 

You receive a citizen complaint about conditions at a store in your community. What fire and life safety concerns exist with the conditions shown?

____

____

____

6. 

This photo shows an outside stair serving bleachers. There is 5 feet of clear width between each handrail. What is the capacity of this stair?

____

____

____

7.

This photo shows bleachers with 14 seats per row with egress available at both ends of each row. What is the minimum spacing between rows (seat back to self-rising seat behind)?

8.

This type of building construction was popular in the early 1900s. These buildings have non-combustible, masonry exterior walls and wooden floor and roof assemblies using standard dimensional lumber. What type of construction is this building?

9.

What fire code violation can be seen on the exit door shown?

10.

The guests at this hotel travel along exterior walkways to open stairways. What section of the means of egress chapter would contain the requirements for protecting this type of arrangement?

11. 

The building under construction uses steel framing for the occupied spaces but has wooden roof trusses. Would this be classified as combustible or non-combustible construction?

____

____

____

12. 

This is an unsprinklered school, the side door from a classroom is located 23 feet from the set of triple doors in the foreground. Can you identify any fire code violations with the triple door assembly shown in this photo?

____

____

____

13. 

What is the construction type for the building shown in the photo?

____

____

____

14. 

The apartment building shown has a brick exterior; is this considered a combustible or non-combustible building?

____

____

____

15. 

Identify the major life safety risk shown in this photo and find the section in your fire or life safety code to address the condition.

16. 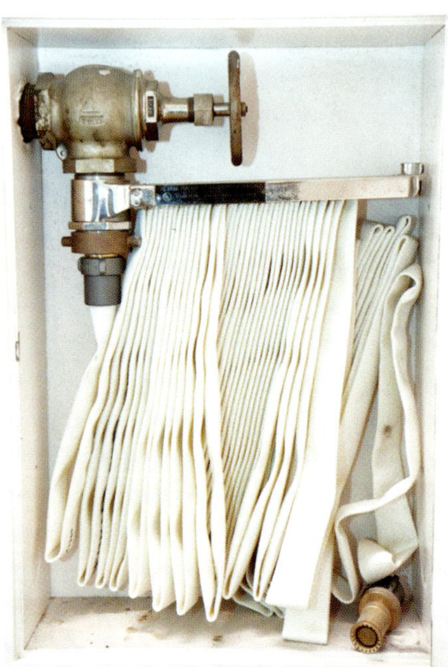

What class of standpipe is shown in this photo?

17. 

This photo shows storage above cabinets in a sprinkler-protected building. What distance must the storage be maintained below the standard pendant spray sprinkler shown? Find the applicable code section that addresses this condition.

____

____

____

18. 

On a routine inspection, you observe this condition. Please describe the fire code violation and find the applicable fire code section that addresses the condition.

____

____

____

19. 

On a routine inspection, you observe the combustible material being stored in the main electrical service panel room. Please describe the fire code violation and find the applicable fire code section that addresses the condition.

____

____

____

20. 

While inspecting an office building, you observe the condition shown. Please describe the fire code violation and find the applicable fire code section that addresses the condition.

____

____

____

21. 

The fire pump test header shown has six 2-1/2-inch outlets. What is the capacity of the fire pump it is connected to?

22. 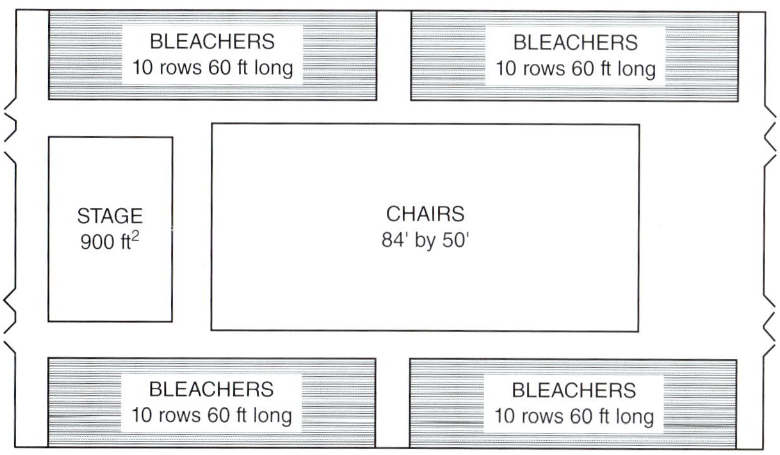

Your local high school wants to have a concert in the gymnasium utilizing the bleachers and chair seating on the floor. There are sixteen 3-ft. wide doors each with a clear opening of 32 inches serving the gymnasium. All doors lead directly to grade level (no steps).

a. What is the total occupant load shown for the bleachers, floor seating, and stage?

b. Is the current egress capacity sufficient for this concert use? What is the capacity of the egress doors?

23. 

The device shown is frequently part of a dry-pipe sprinkler system. What is this device called? What is its purpose?

____

____

____

24. 

A large high-rise hotel in your town has an atrium similar to the one shown in the photo. Describe the fire and life safety risks associated with atriums in buildings. What fire protection features are typically found in buildings with atriums?

____

____

____

Calculate the Occupant Load for a Multiple–Use Occupancy

25.

Calculate the load for a one-story health care facility.

This health care facility has 5 inpatient treatment rooms, 10 sleeping departments, 1 kitchen, and 1 ambulatory health care room.

- Each inpatient treatment room is 20' × 20', with 1 bed 3' × 6.5' and 1 desk 2' × 4'.
- Each sleeping compartment is 10' × 20', with 1 bed 3' × 6.5'.
- The ambulatory care room is 40' × 20', with 4 beds 3' × 6.5'
- The kitchen is 20' × 40'.

a. What is the occupant load of each inpatient treatment room?

____

____

b. What is the occupant load of each sleeping department?

____

____

c. What is the occupant load of the ambulatory health care room?

____

____

d. What is the occupant load of the kitchen?

____

____

e. What is the total occupant load of the building?

____

____

Calculate the Exit Capacity

26. Calculate the exit capacity for each of the components of the means of egress noted for each of the following:

a. An educational classroom building, 100' × 20'

- The corridor is 46" wide
- A stairway is 54" wide
- Doors are 36" wide
- Minimum number of exits required

____

____

b. A non-sprinklered ambulatory health care facility, 40' × 80'

- The corridor is 54" wide
- A stairway is 60" wide
- Doors are 36" wide
- Minimum number of exits required

____

____

c. A board and care facility, 190' × 70'

- The corridor is 48" wide
- The stairway is 60" wide
- Doors are 36" wide
- Minimum number of exits required

____

____

Determine if Plans Meet Egress Code Requirements

27. Review the illustrations. List any variations from Code.

a.

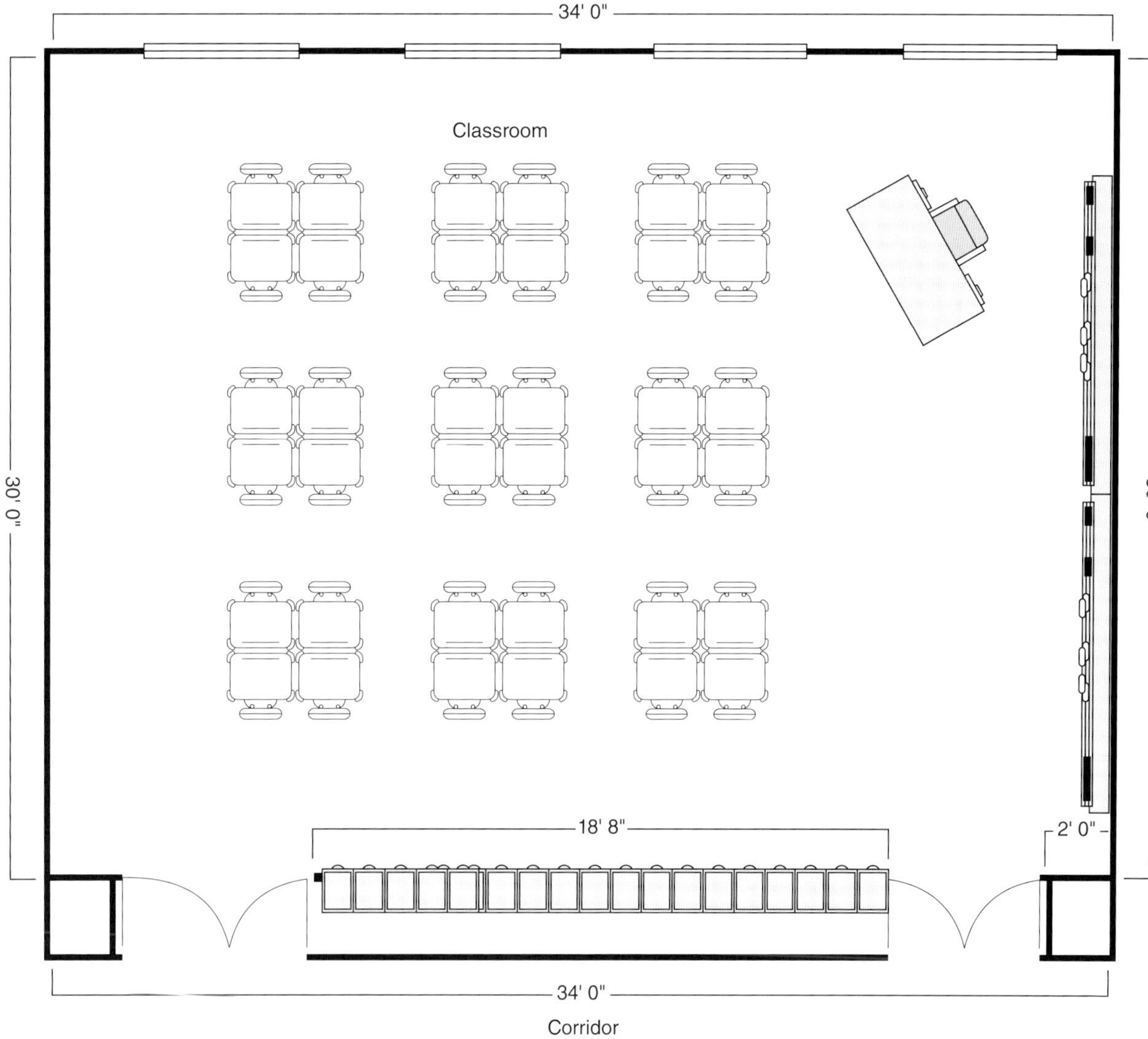

____

____

____

b.

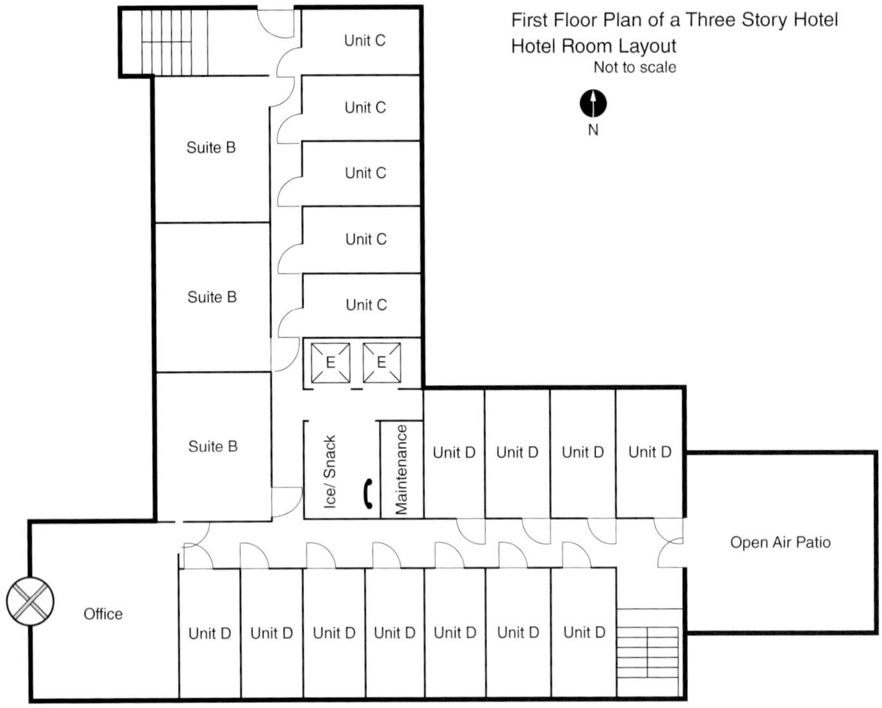
First Floor Plan of a Three Story Hotel
Hotel Room Layout
Not to scale
N
Unit C
Unit C
Unit C
Unit C
Unit C
Suite B
Suite B
Suite B
E
E
Ice/ Snack
Maintenance
Unit D
Unit D
Unit D
Unit D
Open Air Patio
Office
Unit D
Unit D
Unit D
Unit D
Unit D
Unit D
Unit D

c.

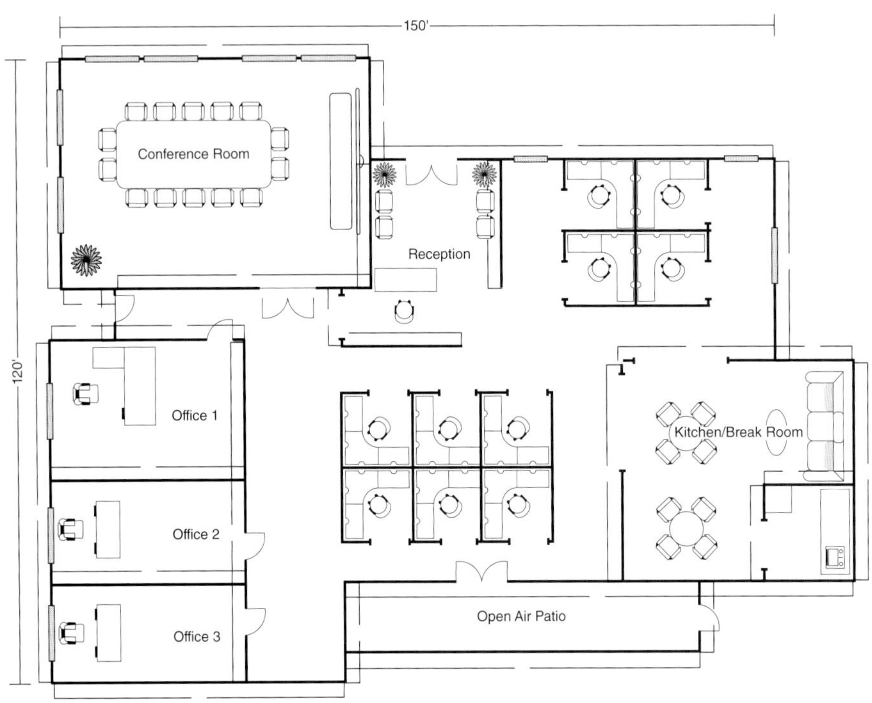
150'
120'
Conference Room
Reception
Office 1
Office 2
Office 3
Kitchen/Break Room
Open Air Patio

d.

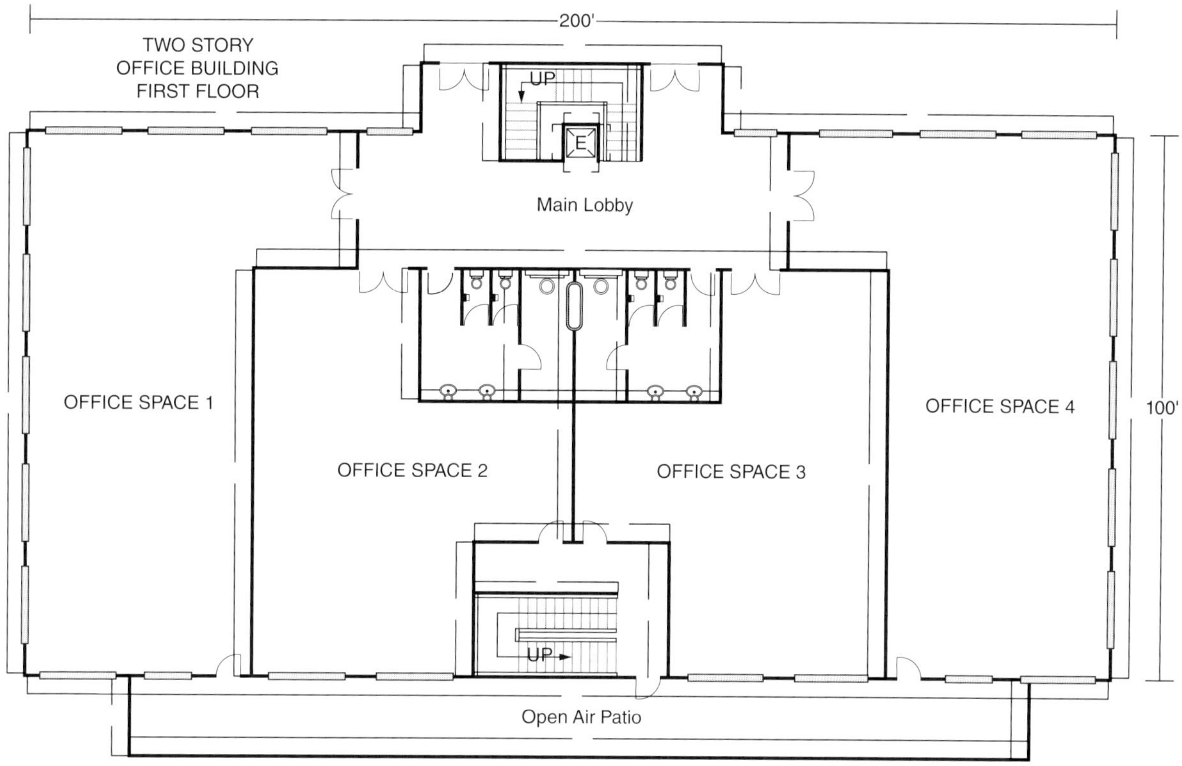
200'
TWO STORY
OFFICE BUILDING
FIRST FLOOR
UP
E
Main Lobby
OFFICE SPACE 1
OFFICE SPACE 2
OFFICE SPACE 3
OFFICE SPACE 4
100'
UP
Open Air Patio

Fire Inspector Level II Answers

1. Shut off valves are not allowed on fire sprinkler systems unless they are an indicating type valve. This shut off valve must be removed as it allows the sprinkler protection to be disabled.

2. It appears that the smoke detector has an unauthorized modification that warrants further investigation of the internal workings of this device. This smoke detector also appears to be dirty and most likely is more than 10 years old requiring replacement per NFPA 72.

3. The pull chain modification is an ON-OFF switch that has been installed to interrupt the battery's power supply to the smoke detector. This switch can render the detector inoperable, which violates the product listing for this device and could endanger the occupants of this residence if the detector is turned off and a fire occurs. If this detector is prone to repeated false alarms, prompting the occupants to make this unauthorized modification, the solution may be to reinstall a detector that has not been tampered or modified away from the source of the nuisance alarms, such as nearby cooking or heating equipment.

4. Combustible storage too close to a heat-producing appliance (furnace). A secondary concern might be the legs installed on the bottom of this furnace. Most manufacturer's instructions recommend that furnaces be installed on a firm base (it appears that this might not be stable and the furnace could move). Consider having a heating contractor verify that the installation is acceptable.

5. There are multiple fire and life safety concerns:
   - Combustible storage is not neat and orderly
   - Merchandise is blocking egress paths and aisles
   - The piles of merchandise do not appear stable (firefighter and occupant safety concern)

6. 3 × 5 ft. = 15 ft. clear width available for exiting.
   15 ft. × 12 inches per foot = 180 inches of clear exit width.
   Stairway Capacity Factor = 0.3 inches per occupant for this occupancy use. (2015 NFPA 101 Table 7.3.3.1; 2012 IBC Section 1005.3.1)
   180 inches/0.3 inches per occupant allows 600 occupants to use this stair.

7. 12 inches. (2015 NFPA 101 Section 13.2.5.5.2; 2012 IBC Section 1028.10.2)

8. Type III construction.

9. Slide bolt locks are not allowed on exit doors.

10. Exit access. (2015 NFPA 101 Section 7.5.3; 2012 IBC 1014.1 and 1019)

11. Combustible; mixed construction gets classified to lower type.

12. Incorrect door swing: at least one of the three doors must swing in the other direction in order to avoid creating a "dead-end" corridor exceeding 20 feet in length on the opposite side of the triple set of doors. (2015 NFPA 101 Table A7.6; 2012 IBC Section 1018.4)

13. Type II-000 (NFPA) or Type II-B (ICC); unprotected steel beam, column, and steel bar joist roof system make it Type II (unprotected).

14. Combustible (Type V).

15. Interior exit stairs are required to be enclosed. Unprotected vertical opening (more than 2 adjacent stories open to each other). (2015 NFPA 101 Sections 7.1.3.2, 7.2.2.5.1.2, 8.6.8, 8.6.9.1; 2012 IBC Sections 707.3.2, 712.1.8, and 1022.1)

16. Class III (characterized by both 2-1/2" and 1-1/2" hose connections).

17. 18 inches. (2013 NFPA 13 Section 8.5.6.1)

18. Exposed wiring on cord. (2015 NFPA 1 Section 11.1; 2015 NFPA 70E Section 205.14(1); 2012 IFC Section 605.1)

19. Combustible storage is prohibited in electrical rooms.

20. Relocatable power taps must be plugged directly into an electrical receptacle. (2015 NFPA 101 Section 11.1.4.2; 2012 IFC Section 605.4.2)

21. The pump capacity falls within the range of 1,250 gpm and 2,000 gpm. (2013 NFPA 20 Sections 4.26, Table 4.26(a))

22. a. 2,260 persons (Bleachers: 60 ft. length/18 inches = 40 people per row. 40 rows × 40 people per row = 1,600 people. Chairs: 84 × 50 = 4,200 sq. ft./7 sq. ft per person = 600 people. Stage: 900 sq. ft/15 sq. ft per person = 60 people. 1,600 + 600 + 60 = 2,260 people).

b. Yes; 2,560 people (16 doors × 32 inches per door = 512 inches. 512 inches/0.2 inches per person = 2,560 people)

23. It is called a drum drip. Its purpose is to allow water and condensation in a dry-pipe sprinkler system to accumulate so it can be safely drained out.

24. Atriums form large unprotected vertical openings that allow smoke and heat travel from lower to upper levels. In addition, hotel guests in this configuration must travel on balconies to reach the enclosed egress stairs; this exposes them to smoke and heat. Another potential risk is the amount of combustibles placed on the lower level. Most buildings with atriums are required to have sprinkler systems, smoke control systems, and fire alarm and detection systems (to operate the smoke control system).

25. a. 400 sq. ft. divided by the occupant load factor of 240 sq. ft., gross area, per occupant equals 1.67, or rounding up, 2 occupants per inpatient treatment room. (Note that this is not 2 patients per room, but 2 occupants per room for the purposes of determining egress calculations. The number of people actually in the room is not restricted until the calculated occupant load exceeds 49 occupants.)

b. 200 sq. ft. divided by 120 sq. ft., gross area, per occupant equals 1.67, or rounding up, 2 occupants per sleeping compartment.

c. The area of the ambulatory health care room is 40 ft. × 20 ft., or 800 sq. ft. 800 sq. ft. divided by 100 sq. ft., gross area, per occupant equals 8 occupants in the ambulatory health care room.

d. The area of the kitchen is 800 sq. ft. 800 sq. ft. divided by 100 sq. ft., gross area, per occupant equals 8 occupants in the kitchen.

e. 5 inpatient treatment rooms, 2 occupants each, equals 10 occupants; 10 sleeping compartments, 2 occupants each, equals 20 occupants; ambulatory health care Room has 8 occupants; kitchen has 8 occupants. The total occupant load for this building is 10 + 20 + 8 + 8 = 46 occupants.

26. a. To calculate the exit capacity of a stairway in an educational facility, you must divide the width of the stairway by the capacity factor from Table 2 on page 124.

$$54 \div 0.3 = 180, \text{ the exit capacity of the stairway}$$

To calculate the exit capacity of a door in an educational facility you must divide the width of the door by the capacity factor from Table 2 on page 124.

$$36 \div 0.2 = 180, \text{ the exit capacity of each door}$$

To calculate the exit capacity of the corridor, you must divide the width of the corridor by the capacity factor from Table 2 on page 124.

$$46 \div 0.3 = 153, \text{ the exit capacity of the corridor}$$

To calculate the minimum number of exits, you must determine the occupant load, then cross check with Table 3 on page 127. Occupant load is the area divided by the occupant load factor from Table 1, page 120–121. Area = Length × Width, 100 × 20 = 2,000 ft^2. 2,000 ÷ 20 = 100, the occupant load. Checking with Table 3, page 127, an occupant load of 100 requires a minimum of 2 exits.

b. To calculate the exit capacity of a stairway in a non-sprinklered health care facility, you must divide the width of the stairway by the capacity factor from Table 2 on page 124.

60 ÷ 0.6 = 100, the exit capacity of the stairway

To calculate the exit capacity of a door in a non-sprinklered health care facility you must divide the width of the door by the capacity factor from Table 2 on page 124.

36 ÷ 0.5 = 72, the exit capacity of the door

To calculate the exit capacity of the corridor, you must divide the width of the corridor by the capacity factor from Table 2 on page 124.

54 ÷ 0.3 = 180, the exit capacity of the corridor

To calculate the minimum number of exits, you must determine the occupant load, then cross check with Table 3 on page 127. Occupant load is the area divided by the occupant load factor from Table 1, page 120–121. Area = Length × Width, 40 × 80 = 3200 ft^2. 3200 ÷ 100 = 32, the occupant load. Checking with Table 3, page 127, an occupant load of 0 to 499 requires a minimum of 2 exits, with limited exceptions.

c. To calculate the exit capacity of a stairway in a board and care facility, you must divide the width of the stairway by the capacity factor from Table 2 on page 124.

60 ÷ 0.4 = 150, the exit capacity of the stairway

To calculate the exit capacity of a door in a board and care facility, you must divide the width of the door by the capacity factor from Table 2 on page 124.

36 ÷ 0.2 = 180, the exit capacity of each door

To calculate the exit capacity of the corridor, you must divide the width of the corridor by the capacity factor from Table 2 on page 124.

48 ÷ 0.3 = 160, the exit capacity of the corridor

To calculate the minimum number of exits, you must determine the occupant load, then cross check with Table 3 on page 127. Occupant load is the area divided by the occupant load factor from Table 1, page 120–121. Area = Length × Width, 190 × 70 = 13,300 ft^2. 13,300 ÷ 200 = 66.5 or 67, the occupant load. Checking with Table 3, page 127, an occupant load of 0 to 499 requires a minimum of 2 exits.

(2015 NFPA 101 Sections 7.3.3, 7.4; 2012 IBC Section 1005.3)

27. a. Occupant load is calculated by determining the size of the room and dividing by the occupant load factor for a classroom:

30 ft. × 34 ft. = 1,020 sq.ft.

1,020 sq.ft./20 sq.ft. per occupant = 51 occupants

Since the occupant load is 50 or more, two exits are required from the classroom and the doors must swing in the direction of travel.

b. Unit and suite doors may swing outward if these doors do not reduce the required width of egress in the corridor by more than 7 inches when fully opened and at no times obstruct more than half of the required exit width in the corridor. (2015 NFPA 101 Sections 7.2.1.4.3.1, 7.2.1.4.3.2; 2012 IBC 1005.7.1)

The Patio must be provided with a gate leading out to the exterior.

The two stairways require separation and enclosure in fire-resistive construction from the floor with exits directly to the exterior that lead to a public way.

The elevator shafts require a fire rated and smoke protected door or elevator lobby.

c. The three exterior exit doors must swing in the direction of egress as they clearly serve a building with an occupant load of more than 49 occupants.

The door from Office 1 cannot reduce the required width of egress in the corridor by more than 7 inches when fully opened and at no times obstruct more than half of the required exit width in the corridor.

d. The three exterior exit doors must swing in the direction of egress as they clearly serve a building with an occupant load of more than 49 occupants.

Photo Credits

Chapter 8
08-01 © AbleStock; 08-02 Courtesy of Fire Fighting Enterprises, Ltd.

Chapter 9
09-08 Courtesy of A. Maurice Jones, Jr.; **09-09** Courtesy of A. Maurice Jones, Jr.; **09-10** Courtesy of A. Maurice Jones, Jr.; **09-12** Courtesy of A. Maurice Jones, Jr.

Chapter 12
12-04 Courtesy of Jack B. Kelley, Inc.; **12-05** Courtesy of Jack B. Kelley, Inc.; **12-06** Courtesy of Jack B. Kelley, Inc.; **12-08** Courtesy of Eurotainer